NEUROSCIENCE RESEARCH PROGRESS

NEUROCYSTICERCOSIS

FROM DIAGNOSIS TO TREATMENT

NEUROSCIENCE RESEARCH PROGRESS

Additional books and e-books in this series can be found on Nova's website under the Series tab.

NEUROSCIENCE RESEARCH PROGRESS

NEUROCYSTICERCOSIS

FROM DIAGNOSIS TO TREATMENT

MARK A. CHAVEZ
EDITOR

Copyright © 2021 by Nova Science Publishers, Inc.

All rights reserved. No part of this book may be reproduced, stored in a retrieval system or transmitted in any form or by any means: electronic, electrostatic, magnetic, tape, mechanical photocopying, recording or otherwise without the written permission of the Publisher.

We have partnered with Copyright Clearance Center to make it easy for you to obtain permissions to reuse content from this publication. Simply navigate to this publication's page on Nova's website and locate the "Get Permission" button below the title description. This button is linked directly to the title's permission page on copyright.com. Alternatively, you can visit copyright.com and search by title, ISBN, or ISSN.

For further questions about using the service on copyright.com, please contact:
Copyright Clearance Center
Phone: +1-(978) 750-8400   Fax: +1-(978) 750-4470   E-mail: info@copyright.com.

NOTICE TO THE READER

The Publisher has taken reasonable care in the preparation of this book, but makes no expressed or implied warranty of any kind and assumes no responsibility for any errors or omissions. No liability is assumed for incidental or consequential damages in connection with or arising out of information contained in this book. The Publisher shall not be liable for any special, consequential, or exemplary damages resulting, in whole or in part, from the readers' use of, or reliance upon, this material. Any parts of this book based on government reports are so indicated and copyright is claimed for those parts to the extent applicable to compilations of such works.

Independent verification should be sought for any data, advice or recommendations contained in this book. In addition, no responsibility is assumed by the Publisher for any injury and/or damage to persons or property arising from any methods, products, instructions, ideas or otherwise contained in this publication.

This publication is designed to provide accurate and authoritative information with regard to the subject matter covered herein. It is sold with the clear understanding that the Publisher is not engaged in rendering legal or any other professional services. If legal or any other expert assistance is required, the services of a competent person should be sought. FROM A DECLARATION OF PARTICIPANTS JOINTLY ADOPTED BY A COMMITTEE OF THE AMERICAN BAR ASSOCIATION AND A COMMITTEE OF PUBLISHERS.

Additional color graphics may be available in the e-book version of this book.

Library of Congress Cataloging-in-Publication Data

ISBN: 978-1-53619-791-4

Published by Nova Science Publishers, Inc. † New York

Contents

Preface		**vii**
Chapter 1	Treatment of Neurocysticercosis: Current and Future Options *Shweta Sinha, Vivek Lal and Rakesh Sehgal*	1
Chapter 2	The Challenge of Deciphering Certainty from Ambiguity for the Laboratory Diagnosis of Neurocysticercosis *Rimanpreet Kaur, Naina Arora, Suraj Singh Rawat, Anand Kumar Keshri, Neha Singh, Avinash Singh, Shweta Tripathi and Amit Prasad*	43
Chapter 3	Trends in the Diagnosis of Human Neurocysticercosis: Issues and Challenges *Abhishek Mewara and Nancy Malla*	73
Index		**101**

PREFACE

Neurocysticercosis, a preventable parasitic infection of the central nervous system caused by tapeworm, is a serious, potentially fatal disease that can cause neurologic syndromes such as epileptic seizures. Chapter one of this monograph describes the existing treatment options for neurocysticercosis, along with possible therapeutic alternatives under different drug development phases. Chapter two explains the difficulties associated with correctly diagnosing neurocysticercosis, resulting from its varied clinical presentation, and mentions current guidelines of diagnostic criteria for neurocysticercosis. Chapter three describes the various merits and demerits of techniques for detecting the parasites associated with neurocysticercosis, which include radioimaging, genotyping of cysts, and antibody, antigen, and nucleic acid detection in body fluids.

Chapter 1 - Neurocysticercosis (NCC), caused by the metacestode (larva) of the tapeworm *Taenia solium* is one of the common parasitic infections of the central nervous system mainly flourish in subarachnoid space, ventricles and, cerebral parenchyma, and is one of the leading causes of acquired epilepsy. It has become a serious public health issue, mainly in countries of developing economies such as Latin America, Africa, China, and India, and presently it is also a major public health concern in the United States. Neuroimaging is the main diagnostic test for NCC, apart from several immunological and molecular tests. Therapeutic measures include symptomatic therapy, antiparasitic drugs (albendazole and praziquantel), and with or without surgery. Albendazole and praziquantel are capable of removing the cysticerci, however, proven cure rates are low and required prolonged anti-parasitic therapy mainly in case of complicated disease. NCC management is still under debate mainly with respect to its uses and benefits achieved from these cysticidal drugs. Potential drug targets such as proteases, kinases, G-protein-coupled receptors, and a few more can be exploited for the sake of novel drug development. This chapter will elaborate about the existing treatment options for NCC, along with possible therapeutics alternatives which are under different drug development phases.

Chapter 2 - Neurocysticercosis (NCC) is a major cause of acquired epilepsy in developing regions of the world where socio-economic indexes are low and hygiene and health awareness is less among the communities. According to a WHO report, NCC causes approximately 50,000 human deaths per year, and in 2014 the parasite *T. solium* was ranked first on global scale of food born parasites. The WHO Food Borne Disease Burden Epidemiology Reference Group 2015 had identified "*T. solium* as a leading cause of deaths by food borne diseases considerable to 2.8 million disability adjusted life years loss (DALYs)." Another recent study had estimated that NCC accounts for at least 5% of all avoidable epilepsy cases across globe and approximately 30% of new acquired epilepsy cases in endemic areas. In a WHO report of 2015, approximately 0.317%, 0.076% and 0.597% (highest) of total population is estimated to be at risk of infection in Peru, Ecuador and India, respectively. Hence, for endemic areas it is a major cause of active epilepsy. In Global Burden of Disease Study (GBDS) 2010 report, it was estimated that 0.07 DALYs are lost per 1000 people globally due to occurrence of NCC, but still this data was considered to be an under estimation due to complexity associated with the accurate diagnosis of NCC. The diagnosis of NCC is complex due to its varied clinical presentation, which ranges from mild headache to severe recurring seizure; and in some extreme rare cases mortality is also reported. The clinical manifestations of NCC varies from person to person due to differences in the load and stage of the loaded parasite in the person.

This extreme variation in clinical outcome of infection also depends upon several environmental and host genetic/immunological factors which still need to be identified accurately. The most significant clinical sign is occurrence of acute symptomatic multiple episodes of seizures which is observed in 80% of cases; the other conditions described in symptomatic infections are headache, chronic meningitis, focal neurological deficit hydrocephalus, spinal and ocular cysts, increased intracranial pressure or cognitive decline, psychological disorders etc.. Considering the pleiomorphic clinical presentation and difficulty in diagnosis of NCC, a detailed guideline was made by International Working Group on NCC of International League Against Epilepsy (ILAE) in 2004 that considered clinical, neuroimaging, immunological and epidemiological data to identify the definitive or probable cases of NCC. The guideline was revised again in 2017 to make it simple, these guidelines are mentioned below.

Chapter 3 - Human neurocysticercosis (NCC) is a disease caused when the larval stage (cysticerci) of *Taenia solium* lodge in the central nervous system. It is regarded as a significant public health problem in Asia, Africa and the Latin America. The infection is also increasingly being seen in more developed countries due to frequent travel and immigration from endemic areas. Taeniasis/cysticercosis is one of the neglected tropical diseases which are targeted for control by the WHO. The clinical diagnosis of NCC is presumptive and is usually substantiated by laboratory diagnostic procedures. Radioimaging provides information regarding the number, size, and location of the cysts, but often is non-specific, mimicking other pathologies. Antibody, antigen, nucleic acid detection in body fluids and/or genotyping of the cysts usually substantiates and/or confirms the clinical diagnosis. However, all these techniques have their own merits and demerits, with varying sensitivity and specificity which depend upon the sample, assay and type of antigen used, besides clinical presentation including stage and location of the lesion. Therefore, application of a single conventional technique may not provide confirmatory diagnosis in all the clinically suspected patients. Moreover, molecular techniques are not usually available in endemic areas with limited resources and facilities. The combination of two or more techniques, when applied, may yield desirable sensitivity and specificity for the diagnosis and planning of treatment strategies. The detection of antibody response to crude soluble extract *T. solium* antigen by ELISA to achieve desirable sensitivity, followed by more specific technique with the use of lower molecular mass specific antigens to check specificity in seropositive individuals may serve useful purpose for diagnosis in endemic areas which have limited resources for molecular techniques. The diagnostic criteria for human NCC need to be defined specific to the endemic area, depending upon the clinical presentation, asymptomatic infections, clinically similar CNS pathologies and other associated factors in the specific regions. The antigen detection technique(s) in body fluids have been found helpful to assess the response to treatment on follow-up of patients. The advanced molecular techniques, nucleic acid detection by PCR and its modifications, isothermal nucleic acid amplification techniques such as LAMP, cell mediated methods, and advanced proteomics techniques are still under investigation and are yet to find a practical use in the diagnosis of NCC.

In: Neurocysticercosis
Editor: Mark A. Chavez

ISBN: 978-1-53619-791-4
© 2021 Nova Science Publishers, Inc.

Chapter 1

TREATMENT OF NEUROCYSTICERCOSIS: CURRENT AND FUTURE OPTIONS

Shweta Sinha[1], PhD, Vivek Lal[2], MD and Rakesh Sehgal[1,], MD*

[1]Department of Medical Parasitology, Postgraduate Institute of Medical Education and Research, Chandigarh, India
[2]Department of Neurology, Postgraduate Institute of Medical Education and Research, Chandigarh, India

ABSTRACT

Neurocysticercosis (NCC), caused by the metacestode (larva) of the tapeworm *Taenia solium* is one of the common parasitic infections of the central nervous system mainly flourish in subarachnoid space, ventricles and, cerebral parenchyma, and is one of the leading causes of acquired epilepsy. It has become a serious public health issue, mainly in countries of developing economies such as Latin America, Africa, China, and India, and presently it is also a major public health concern in the United

---

[*] Corresponding Author's E-mail: sehgalpgi@gmail.com.

States. Neuroimaging is the main diagnostic test for NCC, apart from several immunological and molecular tests. Therapeutic measures include symptomatic therapy, antiparasitic drugs (albendazole and praziquantel), and with or without surgery. Albendazole and praziquantel are capable of removing the cysticerci, however, proven cure rates are low and required prolonged anti-parasitic therapy mainly in case of complicated disease. NCC management is still under debate mainly with respect to its uses and benefits achieved from these cysticidal drugs. Potential drug targets such as proteases, kinases, G-protein-coupled receptors, and a few more can be exploited for the sake of novel drug development. This chapter will elaborate about the existing treatment options for NCC, along with possible therapeutics alternatives which are under different drug development phases.

Keywords: Taenia, drug targets, cysticerci

INTRODUCTION

Neurocysticercosis (NCC), caused by the metacestode (larva) of the tapeworm *Taenia solium* is one of the common parasitic infections of the central nervous system (CNS) mainly flourish in subarachnoid space, ventricles and, cerebral parenchyma, and is one of the leading causes of acquired epilepsy and seizures. These larvae or cysticerci may also establish itself in various other tissues like cardiac, eyes, skeletal muscles, tongue, diaphragm, and subcutaneous tissues, which leads to a condition called as 'cysticercosis' (Smith JL, 1994; Esquivel-Velazquez, 2011). NCC is very common in low and middle-income countries where humans and pigs live in close quarters. Most of South and Central America, Southeast Asia, the Indian subcontinent, China, and sub-Saharan Africa are endemic to it. It affects about 50 million of the human population all over the world (Winkler AS, 2012; Delgado-García, 2019), having a similar incidence of disease occurrence in both male and female and mostly a highest among age groups in between 30-40 years (Abba et al. 2010). Although transmission is absent or uncommon in the United States and most of Europe, an increased incidence of NCC has been recorded in recent years in America, owing to a rise in immigration from endemic regions as well as improved diagnostic ease with imaging techniques (Tellez-Zenteno and Hernandez-Ronquillo, 2017). In the United States, it has been reported for approximately 1,800 NCC-associated hospitalizations each year. Also, the overall cost of hospitalization for cysticercosis is on the higher side as compared to malaria and after integrating all other neglected tropical diseases (O'Neal and Flecker, 2015). NCC is usually represented by varied clinical features, and these differences are mainly due to variations in location, number, size, and stages of parasites present in the human nervous system. The variation in clinical features poses a severe impact on the prognosis and treatment of the disease (Rodrigue et al. 2012). Symptomatic seizures and epilepsy are considered to be the most common manifestations of NCC and have been confirmed to be the cause of 20% to 70% of cases of symptomatic epilepsy in some countries (Sharma et al. 2015). In Asian countries, it is responsible for about 50% of epilepsy cases and about 90% of symptomatic seizure cases in children (Singhi P, 2011). Anatomical cyst localization, environmental factors, host biology, parasite infective capacity, and, most importantly, host immune responses all play a role in the production of symptoms (Esquivel-Velazquez, 2011), and because of this non-specific and varying clinical appearance, it is difficult to diagnose clinically. Current therapeutic measures include symptomatic therapy, antiparasitic drugs, and surgery. Among antiparasitic drugs, benzimidazoles and praziquantel are widely used (Del Brutto OH, 2020), however, surgical intervention is mostly done in selective cases. The antiparasitic drugs that are used for NCC treatment have considerable toxicity and provide sub-optimum results as these can only hinder the reproduction and development of parasites in spite of killing them (Brunetti and White 2012). Thus, novel drug targets and compound groups with low toxicity and improved efficacy even at low doses are urgently required for disease prevention, control, and intervention. This chapter will elaborate about the existing treatment options for NCC, along with possible therapeutics alternatives which are under different drug development phases.

LIFE CYCLE

T. solium is a zoonotic parasite having a complex life cycle accompanying two-host (Figure 1). This includes humans as the only definitive host and possess the adult tapeworm, while humans, as well as pigs both, can be intermediate host that carry the larvae or cysticerci. Humans accidentally acquire infection of *T. solium* after consuming raw or partially cooked pork meat containing the larvae stage of the *Taenia*, which eventually mature into the adult stage. This maturation process takes place in the small intestine of a human. Infected individuals discharge the proglottids, containing the eggs of *T. solium* in their feces. When an infected individual defecates in the open environment, then these eggs are carried away by various sources such as wind, water, wind, insects, or by animal feet etc., which contaminate the animal herbage, and this way cattle get infected while consuming it. Pigs, as coprophagic animals, typically get infected by consuming human feces directly or indirectly from polluted habitats. After ingestion, the eggs release hexacanth embryos or oncospheres in the gastrointestinal tract.

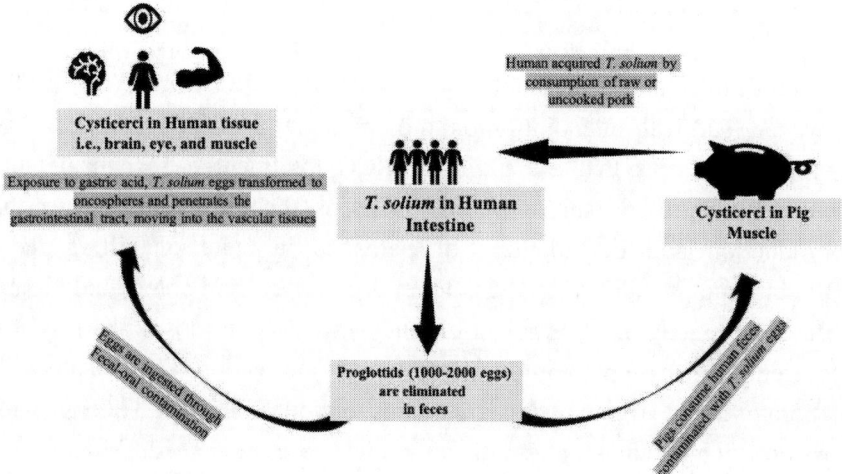

Figure 1. Life cycle of *T. solium*.

Oncospheres then reaches to the bloodstream after penetrating the wall of the small intestinal and there it matures into metacestodes, especially in host tissue such as striated muscles or the CNS (based on the *Taenia sp.*) Metacestodes afterward transform into cysticerci, which are mainly vesicles with an invaginated scolex (Hoberg EP, 2002). The total time span of a *T. solium* is difficult to estimate, but it is most likely just a few years (Lightowlers MW, 2013) and varies significantly from that of *T. saginata* which has a longer life (Ito et al. 2013). The adult *T. solium* has immense biotic potential, therefore the tapeworm carrier should be the key goal for control measures (Del Brutto and Garcia, 2014; Symeonidou et al. 2018).

Clinical Manifestations of NCC

The NCC patient shows unspecified clinical features ranging from asymptomatic, mild symptomatic to life-threatening. Cysticerci inside the CNS affect the parenchyma, subarachnoid space, or intraventricular system and less commonly the ocular and spinal region. Usually, the clinical manifestations are said to be pleomorphic due to disparity in the number, size, location, and stage of the cysts during disease presentation. NCC often leads to adult-onset epilepsy mainly in endemic areas of the world, especially in Latin America, Africa, and Asia (Commission on Tropical Disease of the International League Against Epilepsy, 1994). Generalized tonic-clonic or simple partial seizures are the most common forms of seizures. Patients with the parenchymal disorder are more likely to develop epilepsy, while cysts in the cortical sulci may also cause it (Nash and Garcia, 2011). Seizures caused by cysticercosis are most common when the dying cyst triggers an inflammatory response, but they have also been documented in the cystic stage. For several cases, epilepsy is the only symptom they have, with 50-70% of patients reporting recurring seizures (Budke et al. 2009).

PATHOLOGY OF NCC

Adult *Taenia* expels proglottids (contains almost 1000 - 2000 eggs per proglottids) and after accidental ingestion most often these hexacanth embryones get inside the brain parenchyma through blood circulation and develops into cysticerci. These cysticerci thereafter undergo four different stages of transformation which take place within the time span of few weeks to many years (Escobar A, 1983). These are: (i) vesicular; (ii) colloidal; (iii) granular-nodular, and (iv) calcific stages. The first vesicular stage is distinguished by a cyst with a translucent vesicular wall, clear fluid, and a viable invaginated scolex. There isn't much of an aggressive response from the host at this stage. The vesicular or live stage usually does not provoke any symptoms. During the next step, which is a colloidal stage, the cyst grows a thick vesicular wall, the fluid gets turbid, and the scolex degenerates. At this stage, the pathology shows varying degrees of acute and chronic inflammation, suggesting an active inflammatory response (Escobar A, 1983). Seizures are normal, and radiographic analysis shows cystic lesions with edema and enhancement (Garcia and Del Brutto, 2003). As the cyst progresses to the granular level, it develops a thick vesicular wall, a degenerated scolex, gliosis, and has minimum host inflammatory response. Afterward, the parasite eventually transforms into coarse calcified nodules, which is known as the calcific stage (Del Brutto et al. 1998). The other three stages of the parenchymal cyst's degeneration can result in seizures, but the colloidal stage rarely produces symptomatic mass lesion having raised intracranial pressure (ICP) with or without focal neurological symptoms and signs. When there are many cysts inside the brain, their distortion can cause extreme edema, a high ICP which causes headache, modified sensorium, vomiting, and even death which is termed as "cysticercotic encephalitis" a form of NCC. The majority of NCC patients in Latin America and other endemic countries have multiple cysts in their brains, while approximately two-thirds of patients in India pose mainly a single degenerating cyst called the "solitary cysticercus granuloma (SCG)".

EXTRAPARENCHYMAL NCC

In this form of NCC, cysticerci can reside in the ventricular system or subarachnoid space, resulting in potentially lethal acute ICP. In this case, the flow of cerebrospinal fluid is obstructed due to the presence of cysts that lodged in the fourth ventricle, resulting in hydrocephalus (Nash and Garcia, 2011; Fogang et al. 2015).

OTHER FORMS OF NCC

This includes cysticercal encephalitis, manifest intracranial hypertension, which is caused by a major parenchymal infection that triggers a strong immune response that results in diffuse brain edema (Winkler AS, 2013; Takayanagui et al. 2011). Spinal cord participation is unusual, accounting for just 1%–5% of all NCC cases.

EXISTING TREATMENT

Since NCC is a pleomorphic disorder, it is impossible to create a single treatment plan for all patients. Therefore, chemotherapeutic management is determined by looking at the general patient status (nutrition, immunologic condition), the number, and position of the parasites. (White AC, 2000). Current therapeutic measures include symptomatic therapy, antiparasitic drugs, and surgery (Table 1).

Table 1. Recommended Treatment for different form of NCC (White AC et al. 2018)

Parenchymal NCC	
FORM	**Recommended Treatment**
Viable parenchymal	Antiparasitic drugs should be used every patient except in case of increased of intracranial pressure.
Viable cysts (1-2)	Monotherapy with albendazole (15 mg/kg/d in 2 daily doses up to 1200 mg/d) with food for 10 d
Viable cysts (>2)	Albendazole (15 mg/kg/d in 2 daily doses up to 1200 mg/d) combined with praziquantel (15 mg/kg/d in 3 daily doses) for 10 d Corticosteroids should be used whenever antiparasitic drugs are used Antiepileptic drugs should be used in all patients with seizures
Degenerating (colloidal) cyst Single enhancing lesion due to neurocysticercosis	Albendazole (15 mg/kg/d in 2 daily doses up to 800 mg/d) for 1–2 wk. -Corticosteroids should be given concomitantly with antiparasitic agents. -Antiepileptic drugs should be used in all patients with seizures
Calcified parenchymal neurocysticercosis with or without perilesional edema	Antiparasitic treatment not recommended Treatment with antiepileptic drugs Corticosteroids should not be routinely used
Cysticercal encephalitis (with diffuse cerebral edema)	Avoid antiparasitic drugs, treat diffuse cerebral edema with corticosteroids
Extraparenchymal NCC	
Intraventricular (lateral or third ventricle	Removal of the cysticerci by minimally invasive, neuroendoscopy when feasible
Intraventricular (fourth ventricle)	Either endoscopic or microsurgical cystectomy is suitable, depending on the experience of the surgeon
Intraventricular—when surgical removal not feasible (e. g., adherent cyst)	CSF diversion via a ventriculoperitoneal shunt Adjuvant antiparasitic and anti-inflammatory therapy
Subarachnoid -Giant cyst (usually in Sylvian fissure) Basal subarachnoid (racemose)	Surgical excision or antiparasitic treatment Albendazole 15mg/kg/day for more than 1month, together with high doses of steroids Albendazole 15 mg/kg/day for more than one month, together with high doses of steroids
Spinal NCC	Intramedullary/extramedullary spinal cysticercosis should be surgical -Corticosteroid treatment of patients with Spinal NCC with evidence of spinal cord dysfunction (e.g., paraparesis or incontinence) or as adjunctive therapy along with antiparasitic therapy (strong, moderate)

Symptomatic Therapy

Corticosteroids

Corticosteroids are recommended as a first-line treatment for the management of inflammatory reaction that arises in the course of natural healing of the lesions, or because of the resultant adverse effect of the start-up therapy in NCC patients. The usual regimen includes dexamethasone at dosages of somewhere in the range of 4.5 and 12 mg/day (García et al. 2002). Dexamethasone can be replaced by prednisone at the dosage of 1 mg/kg/day in a condition where prolonged treatment with steroid is required. Corticosteroids are considered as the immediate treatment option for chronic cysticercosis arachnoiditis or encephalitis. Daily dosages of up to 32 mg of dexamethasone are suggested for reducing the brain edema accompanying this condition (Del Brutto et al. 1993). In case of acute intracranial hypertension as a consequence of NCC, mannitol, at the dosages of 2 g/kg/day, can be given (García et al. 2002). However, till now there is not a single unified regimen or standard of use (Cuello-García et al. 2013), as per the latest advice in clinical practice guidelines (AC White et al. 2018).

Antiepileptic Drugs (AEDs)

AEDs play the crucial role in the management of seizures, a common manifestation in patients with NCC. It helps in preventing the frequency of seizures in symptomatic epilepsy patients, an outcome of NCC. It can also help the NCC patients who have some other clinical features apart from seizures such as altered behavior or headache (Frackowiak et al. 2019). First-line AEDs such as carbamazepine or phenytoin are adequate in handling seizures (Frackowiak et al. 2019). In the case of NCC, it is yet not clear about the time duration of AEDs uses particularly after having an acute episode of seizure, usually a host immune system-conciliated inflammatory response raised against degenerating cysticerci. Mainly it is recommended to carry on with the treatment till the lesions in the brain gets clear up, which can be determined through neuroimaging procedure. In clinical practice, repeated CT scan or MRI is preferred within a period of 6 months to confirm the full-length treatment, and accordingly, AED medication can be suggested for 12 weeks' time period depending upon improvement in the lesions. In case of unresolved seizures, which may occur due to inactive calcified cysts, treatment with AED usually suggested continuing till there is no occurrence of seizure for about 2 years (Sinha and Sharma 2009). According to experts, most patients with NCC seizures, especially those with SCG and solitary calcification, counter well to monotherapy with a first-line AED (Bustos et al. 2016).

ANTI-PARASITIC/CYSTICIDAL THERAPY

Cysticidal drugs are crucial as it has increased the prognosis of several patients suffering from NCC. Two most efficacious and widely used cysticidal drugs against *T. solium* cysticerci are, praziquantel and albendazole (Del Brutto OH, 2020).

Praziquantel: It is an acylated isoquinoline-pyranzine that was introduced in the year 1972 by Bayer and Merck. It has a wide spectrum antihelminthic action that acts on several intestinal cestodes and is highly active against *T. solium* cysticerci harboring in the tissues such as the brain, muscles, eye, and others. Since its discovery, it has taken over many drugs as the sole treatment option for several helminthic infections. It is administered orally, is safe, and highly effective. Praziquantel taken orally goes through rapid first-pass hepatic metabolism. The peak serum concentration reaches within 1 to 2 hours with a short elimination half-life which comes in between 1 to 3 hours. The drug concentration often increased when it concomitantly administered with carbohydrate rich diet. Praziquantel is permeable to blood-brain barrier which explains its role in the treatment of parenchymal brain cysticercosis (Sotelo and Jung 1998). The clearance rate of parenchymal cysts is almost 60–70% after following a 15-days treatment schedule of praziquantel at a dose of 50 mg/kg/day

(Nash TE 2003; Sotelo et al. 1984, 1985). Currently, a new regimen of single-day treatment with praziquantel has been suggested, which shows a similar outcome as a long-term treatment. This regime includes administration of 25/mg/kg thrice a day at intervals of 2 hours, which has a basis to enhance the time exposure of parasite to elevated drug concentrations. This regime has also added advantages of cutting the cost of medication, shorting the length of treatment and the total received dose reduction (cutting to about one-tenth of the original dosage) (Corona et al. 1996; Del Brutto et al. 1999; López-Gómez et al. 2001). When combined with AEDs or corticosteroids, the drug's bioavailability is significantly reduced, such as carbamazepine, phenytoin, and dexamethasone, in particular (Sotelo and Jung 1998).

The mechanism by which praziquantel exerts its effect is still not properly illustrated, however, few studies depict that praziquantel enhances the membrane permeability to calcium ions, leading to paralysis and damage of the tegument. The incapacitated parasite loses its integrity and displace from its location and goes inside circulation, where it is attacked by the various components of the host immune system, such as phagocytes, natural killer cells, etc., (Overbosh et al. 1987). Other mechanisms of action by which the drug exerts its effect, include focal disintegrations. Recently a new mechanism of action is hypothesized, that reveals blockage of adenosine uptake under drug influence which ultimately leads to parasites death as there is the absence of functional de novo purine biosynthesis pathways in these cestodes. The most common side effects include disturbances in CNS (malaise, headache, and dizziness); Gastrointestinal tract (GIT) disturbances (abdominal cramps, nausea, diarrhea, and colitis), and allergic reactions (rashes, urticaria, eosinophilia, and pruritus). These side effects are mainly a result of pronounced immune system in response to dying parasites and its released components (Ahmad et al. 2017).

Albendazole: It belongs to a class of benzimidazoles, with a wide spectrum of antihelminthic action. It is used for the control and treatment of NCC and is observed to be more efficacious as compared to praziquantel (Sotelo et al. 1988a, b, 1990; Takayanagui and Jardim 1992).

Albendazole is not properly absorbed through the GIT, mainly because of its poor aqueous dissolution but can quickly metabolize to metabolite i.e., albendazole sulfoxide, constituent responsible for anthelminthic activity. Prior albendazole was recommended to be given for 1 month at a dosage of 15 mg/kg/day, which is changed to a 7-day course at the same dosages. This short course treatment is found to be superior to praziquantel treatment (García et al. 1997). Additionally, albendazole can also demolish subarachnoid and ventricular cysts (Del Brutto OH, 1997). However, in few cases, especially in patients that have huge subarachnoid cysts, doses (as high as up to 30 mg/kg/day) or for the increased duration, or even repeated, courses of albendazole may be suggested. In parenchymal NCC, sole treatment with albendazole can be given if patients have one or two cysts, however, a combined treatment of praziquantel and albendazole can be chosen for patients with more than two cystic lesions (Del Brutto OH, 2020). The advised dosage regimen is mostly 15 mg/kg/day given for 8-30 days (Garcia et al. 2016).

Albendazole exerts its antihelminthic effect by hindering the process of cell division, which occurs in the intestinal cells of the adult worm. The drug bind to the colchicine-sensitive site of tubulin, thereby obstructing its polymerization and formation into microtubules. As a consequence of this disintegration in the microtubular assembly, the glucose uptake of the parasites gets impaired, which results in depletion of their glycogen stores and starvation. This further causes impairment of the mitochondria and endoplasmic reticulum of the germinal layer and then the liberation of lysosomes results in parasite death (Rossignol JF, 1981). The most usual side effects of albendazole treatment include stomach pain, headache, dizziness, and alopecia. (Cruz et al. 1995).

Correct administration of cysticidal drugs is a subject of dilemma mainly because of the asymptomatic nature of NCC which often raises complexity in choosing correct management for NCC patients (Del Brutto OH, 2012). Usually, cysticidals are effective and safe, and their side effects are the results of parasite killing and involvement of the host-immune system during this process, rather than drug toxicity. However, monitor checks such as blood cell count and liver function tests are advisable in patients who are on a prolonged regimen of albendazole, as gradually it may lead to pancytopenia and hepatotoxicity (Choi et al. 2008; Opatrny et al. 2005). Importantly, patients having cysticercotic encephalitis, hydrocephalus, ICP arises from other causes (like major non-encephalitic infections, giant subarachnoid cysts), or subretinal cysticercosis (fundoscopic examination is needed before initiation of therapy) should not get cysticidal drugs until these issues have been resolved (Del Brutto OH, 2020). The administration of cysticidal drugs should be personalized, in patients showing ventricular cysts. Although albendazole significantly disrupts ventricular cysts, if the cysts are inside the fourth ventricle or close to the foramina of Monro, the inflammatory reaction can trigger acute hydrocephalus. Additionally, patients showing only calcifications should not get cysticidal drugs, as appeared lesions are the outcome of killed parasites (Garcia et al., 2002; Del Brutto OH. 2012). Moreover, the consequences of these anthelminthic drugs on the host immune response are still not validated properly (Prodjinotho et al. 2020).

Current guidelines (White AC et al. 2018), suggest albendazole as a 10-days regime at the dosages of 15 mg/kg/day (equivalent to 1,200 mg per day) for the treatment of patients having one or two parenchymal brain cysts (vesicular or colloidal stage). In the case of patients who have greater than two cysts, it is advisable to take praziquantel at a dose of 50 mg/kg daily for a period of 10-14 days. Cysticidals are not recommended during pregnancy. Concomitant administration of corticosteroids is suggested to most patients, while AEDs are recommended to those who had previous tales of seizures or epilepsy. Repetitive treatment with the same regime should be given in case of persistent cyst visualized on MRI after six months following initial therapy. In subarachnoid NCC, small cysts are located deep inside the cortical sulci and look like as parenchymal cysts. In this case, similar treatment recommendations are followed as parenchymal NCC (Del Brutto OH. 2020).

Combination regimen: Praziquantel and albendazole as combined administration appear to be an appealing choice since both have a distinct mechanism of action on the parasite. Also, praziquantel in combination improves the bioavailability of albendazole sulfoxide giving an additive effect in treatment (Cobo et al. 1998; Garcia et al. 2011). Nonetheless, future clinical trials are needed in this direction to reach a decisive conclusion (Bygott et al. 2009; Panic et al. 2014).

Interaction of AEDs with cysticidal drugs: Epilepsy caused by viable parenchymal brain cysticerci, in this case, AEDs must be used in conjunction with cysticidal medications. In this situation, it is important to consider the possibility of drug reactions (Romo et al. 2014). Carbamazepine, phenobarbital, and phenytoin, all substantially lower the active metabolite, albendazole sulphoxide levels in the blood, which is believed to be caused by upregulation of CYP3A, an enzyme involved in albendazole sulfoxidation (Lanchote et al. 2002). Carbamazepine and phenytoin both have also been observed to lower the plasma concentration of praziquantel (Bittencourt et al. 1992). However, none of these studies looked at the impact of cysticidal drugs on AED plasma levels. Furthermore, there are no reports examining potential interactions between newer AEDs and cysticidal medications, such as lamotrigine, levetiracetam, and lacosamide (Bustos et al. 2016).

Interaction of steroids with cysticidal drug: Concomitant administration of praziquantel and steroids, decreases the serum concentration of steroids (Vazquez et al. 1987), however till now the effect of this pharmacological interaction on parasiticidal activity is not clear. Simultaneous praziquantel administration can also lower phenytoin and carbamazepine levels in the blood (Bittencourt et al. 1992). Albendazole penetrates the cerebrospinal fluid more effectively, and its concentrations remain unchanged when administered along with steroids (Jung et al. 1990a; 1990b).

SURGERY

Surgical intervention is rare and is often needed in cases related to intraventricular and subarachnoid NCC. A ventriculoperitoneal shunt is required in the case of hydrocephalous along with albendazole and steroids as concomitant treatment. Steroids are given recurrently to lower the risk for obstructions as well as shunt revisions in the future. During neurosurgical intervention the foremost goal during any of the cases is the removal of cysts and/or minimization of ICP (Winkler AS, 2012). Beside this, surgical intervention is needed for an atypical SCG and in intractable epilepsy related to NCC (Rajshekhar V, 2010).

LIMITATIONS OF APPROVED CYSTICIDAL DRUGS

The above-mentioned cysticidal drugs have been accounted for the treatment of NCC for more than 2 decades, but still, there is a lack of information about definite dosages and time duration for specific clinical conditions. (Kramer LD, 1995). Also, there has been concern about implication and safety of these cysticidal treatments (Singh and Sander 2004; White AC et al. 2018), as prolonged cysticidal treatment may worsen cerebral edema and causes stroke, which may lead to death. Moreover, albendazole is widely used for NCC treatment, but there is a huge variation in the interindividual response, like, few patients need only a single regime of albendazole, others need a repetitive regimen, while some patients remain unresponsive to treatment, which urges the need for new alternatives (Matthaiou et al. 2008). Further, till now there are no new drugs or vaccines in the drug discovery pipeline for NCC treatment, which makes it very crucial for the search of new drugs or drug targets (Ahmad et al. 2017).

REPOSITIONED DRUGS

Available NCC drugs give suboptimal outcomes and may be toxic if the staging, number, and position of CNS lesions have not been carefully determined by neuroimaging. Therefore, it is of prime importance to search and develop new drugs with improved pharmacokinetic properties and efficacy. Strategy of drug repositioning enables identification new application of authorized or investigational drugs which are beyond the range of the earlier medical indication. It has also the added advantages of lower cost in new drug development as a contrast to traditional ways (Pushpakom et al. 2019). However, for future consideration, it is essential to maintain the efficacy and tolerability of these medications.

Nitazoxanide

Nitazoxanide is a parasiticidal agent that works against bacteria, protozoa, nematodes, and trematodes. Nitazoxanide was authorized by the Food and Drug Administration (FDA) in 2002 as an antiprotozoal agent against *Giardia intestinalis* and *Cryptosporidium* species (Singh and Narayan 2011). Its chemical structure is derived from the niclosamide (a taenicide) and was reported to have an action against *T. saginata* and *H. nana* infections in the 1980s (Rossignol and Maisonneuve, 1984). Due to this, nitazoxanide is not considered as a true repurposed medication for taeniasis; however, its high taenicidal activity makes it appealing for off-label use in situations where patients are unresponsive to standard treatment (Vermund et al. 1986; Lateef et al. 2008). After oral administration, nitazoxanide is quickly transformed to tizoxanide, (Broekhuysen et al. 2000), which also shows parasiticidal activity (Stettler et al. 2003). Both nitazoxanide and its metabolite tizoxanide demonstrated a cestocidal effect against *T. crassiceps*, in a time–concentration-dependent manner. Also, both of these drugs were found equally effective as compared to albendazole sulphoxide against *T. crassiceps* cysts under *in vitro* conditions (Palomares-Alonso, 2007). Nitazoxanide found to obstruct the pyruvate ferredoxin oxidoreductase enzyme-dependent anaerobic metabolism, in helminths (Stettler et al. 2003; Walker et al. 2004). However, in the future, more studies are needed to understand its detailed mechanism of action on *T. solium*.

Mepacrine/Quinacrine

Mepacrine is acridine derivative used as an antiprotozoal, antirheumatic, and an intrapleural sclerosing agent. Earlier in 1930s it was accepted as an antimalarial agent to treat malaria (Baird JK, 2011), but superseded by safer and more effective agents (e.g., chloroquine). Later, it was authorized for the therapy of giardiasis (Canete et al. 2006). In protozoans, it has been predicted to act on their cell membrane, mainly a phospholipase A2 inhibitor. However, still there is a gap in understanding the clear mechanism of action against protozoans. It also activates p53 and a known inhibitor of NF-kB and histamine N-methyl transferase. It is found to effective in the case of niclosamide-resistant *T. saginata* infection and considers as a safe, economical, and well tolerable drug (Koul et al. 2000).

Tribendimidine

It is an amidantel derivative and has a wide range of anti-parasitic activity. This drug is reported to be safe and highly effective at its single-dose against several nematodes infections such as *Ascaris, Strongyloides stercoralis* and hookworm in humans and the observed cure rates was 92–96%, 55% and 52–90% respectively (Xiao et al. 2005; Steinmann et al., 2008). The drug is also potent against tapeworms and trematodes (Xiao et al., 2008). Its anticestodal activity has been observed in mice infected with rodent tapeworm (*Hymenolepis microstoma*), in which 50 mg/kg tribendimidine given thrice for three days, has significantly decreased the worm burden by >95% (Kulke et al. 2012). Tribendimidine is an L-subtype nAChR agonist (Hu et al., 2009) and can be promising new anthelmintic. However, till now there is no knowledge of its mechanism of action.

Paromomycin

It is an antimicrobial belong to the class of aminoglycoside. It has a wide spectrum of antibacterial activity. It has been also used in the treatment of various parasitic infections such as amoebiasis, giardiasis, leishmaniasis, and in tapeworm infestation. This drug is authorized in India for the treatment of leishmaniasis (Davidson et al. 2009). Its activity against *Taenia* species was accidentally discovered in the year 1960 in the amoebiasis patient who was co-infected with *T. saginata* (Salem and el-Allaf, 1969). In a follow-up clinical trial, it was observed that daily treatments with 30–50 mg/kg paromomycin for 1–3 days resulted in an 89–100% success rate against *T. saginata*. Furthermore, following a 30 mg/kg single or double dose of paromomycin, an efficacy of approximately 90% was observed against *Hymenolepsis nana* (Salem and el-Allaf, 1969). Paromomycin acts by inhibiting the process of protein synthesis by binding to 16S ribosomal RNA (Panic et al. 2014) in the non-resistant cell.

Tamoxifen

It is a competitive antagonist of the estrogen receptor and used for the treatment of early-stage breast cancer (den Hollander et al. 2013). Tamoxifen supposed to adhere to and interrupt the function of estrogen receptor-like proteins of the parasites which are responsible for crucial physiological processes. Tamoxifen decreases 80 and 50% of parasite burden in female and male mice and causes loss of motility and reduced reproduction when studied under *in vitro* conditions (Vargas-Villavicencio et al. 2007). Tamoxifen also decreases intestinal colonization by *T. solium* by about 70% and recovered one had 80% reduction in their length, which seem as only scolices without strobilar, observed in hamster (Escobedo et al. 2013). However, anti-taeniasis and cysticidal activity of tamoxifen in the case of *T. solium* infected humans and cattle is yet to needed to study (Escobedo et al. 2013).

Metrifonate (Trichlorfon)

It is an organophosphorus compound mainly used as an insecticide and as an anthelmintic agent for animals. It is effective in the treatment of urinary schistosomiasis and onchocerciasis (Holmstedt et al. 1978). Metrifonate is a long-acting cholinesterase inhibitor and inside the body, it is transformed non-enzymatically into 2, 2- dichlorovinyl dimethyl phosphate. In a case report, cysticidal activity of metrifonate was reported (Tschen et al. 1981). Notable improvement was seen in patients of cutaneous cysticerosis with about 75% complete depletion of nodules after administration of two regimens of metrifonate at dosages 10 mg/kg/day for 6 consecutive days at the interval of one month. (Tschen et al. 1981; Panic et al. 2014). The therapy with metrifonate suffer from few side effects such as abdominal cramp, nausea, vomiting and dyspepsia which must be account while its usages.

POTENTIAL/NOVEL DRUG TARGETS

Exploiting structural variations between proteins or enzymes which participates in *T. solium* intermediate metabolism and those of the vertebrate host can be used to identify new chemotherapeutic targets for the treatment of NCC. Therefore, a drug must own some characteristics: non-competitive inhibition, effectiveness against all stages of the parasite's evolution (mainly at *T. solium* cysticerci), radical parasitologic cure (cysticidal effect), oral administration, passage through the blood brain barrier, and efficacy against the entire NCC clinical spectrum (Vielma et al. 2014).

PROTEIN KINASES

Protein kinases are evolutionarily conserved enzymes which participates in the signaling of number of regulatory and developmental processes (Ardito et al. 2017). Kinases' primary function is to transfer phosphate from ATP or GTP to the substrate and cause conformational changes in the substrate, resulting in protein activation. Since the active state of any protein can have a large impact on the physiological function of a cell, kinases are closely operated through phosphorylation, presence of modulator, and substrate inhibitors (Ardito et al., 2017). Due to its implication in disease progression and conserved structure it assumed to be a potential drug target. Recently Arora et al. (2020), has predicted 23 potential kinases to be as a therapeutic target for *T. solium*. Few of them are:

- *Glycogen synthase kinase (GSK):* The GSK kinases is one of the potential drug targets for various diseases. They are involved in regulation of various intracellular metabolic pathway such as glucose regulation. GSK, because of their conserved nature has been illustrated as an optimum drug target for schistosomiasis (Caffrey et al. 2009). In *Taenia* species, the adult worm gets its nutrition in the form of glucose while harboring inside the human, which enable GSK as a probable target for interventions.
- *The Calcium and Calmodulin regulated kinases:* These kinases participate in various crucial physiological processes which are happening inside the cell and are main commander of calcium signaling in health and various diseases. These are stimulated in response to secondary messenger i.e., Calcium (Ca2+) and Calcium sensing protein, Calmodulin, and found to be the second most abundant kinase family in *T. solium* which carries many processes such as muscle contraction, nerve impulse transmission etc. The other key role is in the development of calcareous corpuscles which protect the worms from physiological stress and the host immune attack (Prole and Taylor, 2011; Vargas-Parada and Laclette, 1999). The calcium signaling has significant role in parasite development and its biology, that enables it as a lucrative drug target (Nawaratna et al. 2018). Praziquantel a common wide spectrum antihelminthic agent, is supposed to function by

unbalancing the Calcium homeostasis, but the precise mechanism of action is yet to be validated (Day et al. 2000; Kohn et al. 2001). These kinases may be particularly essential for NCC because the parasite undergoes calcification in later stage of cyst, which leads to epileptogenesis. Moreover, Myosin Light Chain Kinase (MLCK) helps in the formation of myofilaments in the adult worm. These myofilaments organize into bundles to constitute skeleton of the proglottids. The MLCK helps adult worm in muscle contraction during locomotion and passage of immobile molecules, such as glycogen synthase to the germ cells (Willms et al. 2003; Swulius and Waxham, 2008), and it can be another drug target.

- *Tyrosine kinases:* Based on the transmembrane domains, "tyrosine kinases subtypes as receptor tyrosine kinase and cytoplasmic tyrosine kinase". With 34 sequences, it is the third most available kinases in *T. solium*, comprising six receptor kinase families and nine cytoplasmic kinase families. The receptor tyrosine kinases are involved in transmembrane signaling, while cytoplasmic one is responsible for sending signaling to nucleus. These kinases are well-known cancer therapeutic targets as these are flexible in mutation or mostly found in an inactive state (Arora and Scholar, 2005). These kinases have multiple participation in the growth, metabolism and development including neuronal signaling in both vertebrates and invertebrates, making it as a potential target for antihelminthic drugs too (Arora et al. 2020).

PROTEASES

Proteases perform various function in parasitic organisms and contribute to their biology and pathogenesis. The disparate nature of parasite proteases compared to the host orthologous proteins and their immunogenic characteristic has opened opportunities for chemotherapy, as well as for vaccine and serodiagnostic markers development. Yan et al.

(2014), in his study identified 197 novel proteases in *T. solium* which is from 37 families. This data was derived via bioinformatic data mining tool and then differentiated into aspartic, cysteine, serine, metallo-serine, and threonine proteases (Yan et al. 2014), based on the amino acid active site residue (Bos et al. 2009). Prior this, only three proteases were recognized in *T. solium* (Li et al. 2006). Since these proteases are illustrated to perform crucial roles in various process such as, invasion/entry, evasion, virulence, translocation and movement of the parasite inside the host, expression of host immune responses, and regulation of parasite's biology, these newly recognized proteases may stand as potential targets for the creation of new drugs. However, deeper investigation is further needed in this area. Phylogenetic analysis of *T. solium* genome depicts their functional disparity from hosts in case of two regulatory cysteine and serine proteases, therefore these proteases can be a potential druggable targets (Yan et al. 2014; Shareef and Abidi, 2014). Moreover, cysteine protease, i.e., cathepsin L-like peptidase, secreted by Taenia species, found to useful as immunodiagnostic antigen that can be employed for cysticercosis treatment (Zimic M et al. 2009).

Serine proteases: Serine proteases carries out variety of biological processes that includes protein metabolism, blood coagulation, growth control, digestion and fertilization. In *T. solium*, TsAg5 protein was recently identified, as a trypsin-like serine protease with strong homology to the *E. granulosus* antigen Ag5. This TsAg5 obtained from cyst fluid can be further explored as a potential candidate for drug designing and diagnostics (Rueda et al. 2011).

ESTROGEN RECEPTOR-LIKE PROTEINS

Steroidal hormones drastically effect the growth of *T. solium* and *T. crassiceps* (Escobedo et al. 2004, 2010a, b; barra-Coronado et al. 2011; Vargas-Villavicencio et al. 2008) by quelling the viability and reproduction (Escobedo et al. 2004; Vargas-Villavicencio et al. 2008). For example, tamoxifen (an estrogen receptor antagonist) shows protective effect against

T. crassiceps both *in vitro* and *in vivo* (Vargas-Villavicencio et al. 2008). Likewise, RU486 (a progesterone antagonist), hinders the evagination of scolex as well as worm development which is usually induced by progesterone (Escobedo et al. 2010a, b). However, the mechanism by which these steroid hormones affect parasites has not been illustrated completely. Further studies are required to confirm steroidal hormone receptors in *Taenia cysticerci*, which would be helpful in searching and designing of unexplored therapeutic molecules for the treatment of cysticercosis.

INSULIN RECEPTORS

The insulin signaling pathway participates in various cellular processes, such as glucose metabolism and its uptake (Kohn et al. 1996; Ueki et al. 1998; Chang et al. 2004), protein synthesis (Ueki et al. 1998; Oldham et al. 2003), growth and cell division (Chen et al.1996). This signaling pathway is mediated through insulin receptors (IRs) that are responsible for the main development and reproduction in parasites and are usually conserved and widely distributed among helminths (Konrad et al. 2003; Escobedo et al. 2009). The helminths IR was first identified in *Echinococcus multilocularis* (Konrad et al. 2003) and thereafter in *Schistosoma mansoni* (Khayath et al. 2007) and *Schistosoma japonicum* (You et al. 2010). It was observed that insulin of host not only promotes the larval growth but also induces the insulin signaling pathway of *E. multilocularis*, providing a potential novel drug target (Hemer et al. 2014). Nonetheless, very less is known about the function of IRs in *T. solium*. Recently, two IR genes (TsIR-1316 and TsIR-4810) of TsIRs-LBD (ligand binding domains) were identified from *T. solium* genome and these can be evaluated in near future as vaccine or novel drug targets against cysticercosis (Wang et al. 2020).

Acetylcholinesterase's

The enzymatic activity of Acetylcholinesterase's is reduced by antimalarial 'mefloquine' and has been shown to minimize the production of eggs in *E. multilocularis* (VanNassauw et al. 2008). However, due to their low expression in the parasite's developmental stages, acetylcholinesterase's as a target for blocking parasite transmission have limited applicability, but still, it could be another target for drug designing and development.

G-Protein-Coupled Receptors

G protein-coupled receptors (GPCRs) transduce signals from both peptidergic and classical neurotransmitters are of highly valuable for helminth neuromuscular activity. GPCRs covers about 33% prescribed medicines which make them crucial druggable targets that can be explored for helminths infections also (Santos et al. 2017; McVeigh et al. 2018). Also, most of the existing antihelminthics supposed to target the neural communication of the parasite, such as praziquantel which acts on voltage-gated calcium channel subunit and is highly effective on the adult parasite (Marks and Maule, 2010). Tsai et al. (2013) recently classified about 60 putative GPCRs and 31 ligand-gated ion channels, which suggests GPCRs as a valuable future drug target for *Taenia* species also.

Parasite Antioxidant Enzyme

Antioxidant enzymes play a phenomenon role in vital physiological functions and provide protection to parasites from oxidative stress generated by the host during immune evasion. In *T. solium* three enzymes i.e., cytosolic Cu, Zn superoxide dismutase (target of benzimidazoles), a 2-Cys peroxiredoxin, and two isoforms of glutathione transferases (GSTM1 and GSTM2), confined in the larval tegument has been identified which can inactivate host reactive oxygen species produced during immune invasion. Notably, these enzymes remain functional at all developmental stages of the parasite (Vaca-Paniagua et al. 2008), which suggests its critical role in *T. solium* biology and infection and can be explored for novel therapeutic (Ahmad et al. 2017).

PLASMINOGEN AND PLASMIN

T. solium like other helminths such as *S. mansoni* and *Fasciola hepatica* can interplay with the fibrinolytic system proteins mainly with plasminogen and plasmin. This can happen because the parasite either uses or steals the host's proteins in order to avoid being detected by the immune system. The identification of receptors in *Taenia* species that enter plasminogen/or plasmin provides a possible vaccine target that has yet to be validated for cysticercosis (Vielma et al. 2014; Ayón-Núñez et al. 2018).

FUTURE PERSPECTIVES

Despite substantial progress in this area, our understanding of the taeniasis/cysticercosis complex remains minimal, and there are immense challenges to counter especially in endemic areas. The genomes and transcriptomes of tapeworms, which were recently revealed, hold a lot of promise. This strong molecular foundation will aid in the detection of main proteins/enzymes, which will aid in the advancement of diagnosis, vaccine, and therapy (Symeonidou et al. 2018). Moreover, integration of fundamental science with clinical research is required for a better understanding of disease to develop quicker and economical diagnostic methods, and better treatment. Following are the few points that need to be considered or can be worked for better therapeutic outcomes.

1. At 6–12 months of follow-up, corticosteroids were found to diminish the frequency of seizures and fasten the healing of lesions; however, the impact estimation remains uncertain due to risk related to methodological and publication bias. To confirm such evidence, more well-conducted randomized trials comparing the administration of anthelmintics, corticosteroids, and both in combination with placebo are required. (Cuello-García et al. 2013).
2. Heterogeneity of neurocysticercosis: Multidisciplinary investigations in finding heterogeneity at various level (parasite and human) could contribute to better prevention, care, and patient management (Carpio et al. 2013).
3. Studies with uniform methodology performed at multicenter should be adopted to obtain conclusive and reliable data, on which treatment can be decided.
4. A current clinical trial would be very useful in determining which treatment choice is best for patients with subarachnoid NCC.
5. Nanotechnology came out as a promising new therapeutic choice for neurological disorders, which has the probability to transform the approaches of CNS-targeted therapeutics. These nano-engineered molecules can a target particular cell or signaling pathway, can pass the blood-brain barrier, respond to intrinsic stimuli, or can act as drug/gene delivery vehicles, etc. The wide range of nanotechnologies enables the choice of a nanoscale-sized material with the properties that are ideally suited to the therapeutic challenge (Srikanth and Kessler, 2012). Different forms of nanoparticles can be given for cysticidal drugs such as Nanotechnology-based vaccines and immunostimulatory adjuvants, Drug-infused nanoparticles, Nitric oxide-releasing nanoparticles.
6. Detailed studies of *T. soilum's* complex protease-mediated processes and their function in various metabolic activities can aid in a deeper understanding of the infection's establishment in the host as well as helps in designing efficacious drugs.

7. The experimental data on how kinases function in *T. solium* is severely lacking. To specifically associate kinases to their particular function, experimental approaches are required to gather information on the function of kinases in various stages of parasite development which could be further exploited for drug development.
8. Cysteine proteases are another immensely important area for active research for searching novel drug and vaccine targets for cysticercosis.

REFERENCES

[1] Abba, K., Ramaratnam, S. and Ranganathan, L. N. (2010). Anthelmintics for people with neurocysticercosis. *Cochrane database syst rev*, 2010(3): CD000215.

[2] Ahmad, R., Khan, T., Ahmad, B., Misra, A. and Balapure, A. K. (2017). Neurocysticercosis: a review on status in India, management, and current therapeutic interventions. *Parasitol Res*, 116: 21-33.

[3] Ardito, F., Giuliani, M., Perrone, D., Troiano, G. and Lo Muzio, L. (2017). The crucial role of protein phosphorylation in cell signaling and its use as targeted therapy (Review). *Int J Mol Med*, 40(2): 271-280.

[4] Arora, A. and Scholar, E. M. (2005). Role of tyrosine kinase inhibitors in cancer therapy. *J Pharmacol Exp Ther*, 315(3): 971-979.

[5] Arora, N., Raj, A., Anjum, F., Kaur, R., Rawat, S. S., Kumar, R., Tripathi, S., Singh, G. and Prasad, A. (2020). Unveiling Taenia solium kinome profile and its potential for new therapeutic targets. *Expert rev proteomics*, 17(1): 85-94.

[6] Ayón-Núñez, D. A., Fragoso, G., Bobes, R. J. and Laclette, J. P. (2018). Plasminogen-binding proteins as an evasion mechanism of the host's innate immunity in infectious diseases. *Biosci Rep*, 38(5): BSR20180705.

[7] Baird, J. K. (2011). Resistance to chloroquine unhinges vivax malaria therapeutics. *Antimicrob Agents Chemother*, 55: 1827-1830.
[8] barra-Coronado, E. G., Escobedo, G., Nava-Castro, K., Jesús Ramses, C. R., Hernández-Bello, R., García-Varela, M., Ambrosio, J. R., Reynoso-Ducoing, O., Fonseca-Liñán, R., Ortega-Pierres, G., Pavón, L., Hernández, M. E. and Morales-Montor, J. (2011). A helminth cestode parasite express an estrogen-binding protein resembling a classic nuclear estrogen receptor. *Steroids*, 76(10-11): 1149-1159.
[9] Bittencourt, P. R., Gracia, C. M., Martins, R., Fernandes, A. G., Diekmann, H. W. and Jung, W. (1992). Phenytoin and carbamazepine decreased oral bioavailability of praziquantel. *Neurology*, 42(3 Pt 1): 492-496.
[10] Bos, D. H., Mayfield, C. and Minchella, D. J. (2009). Analysis of regulatory protease sequences identified through bioinformatic data mining of the Schistosoma mansoni genome. *BMC Genomics*, 10: 488.
[11] Broekhuysen, J., Stockis, A., Lins, R. L., De Graeve, J. and Rossignol, J. F. (2000). Nitazoxanide: pharmacokinetics and metabolism in man. *Int J Clin Pharmacol Ther*, 38: 387-394.
[12] Budke, C. M., White, A. C. and Garcia, H. H. (2009). Zoonotic larval cestode infections: neglected, neglected tropical diseases? *PLoS Negl Trop Dis*, 3:e319.
[13] Bustos, J. A., García, H. H., and Del Brutto, O. H. (2016). Antiepileptic drug therapy and recommendations for withdrawal in patients with seizures and epilepsy due to neurocysticercosis. *Expert Rev Neurother*, 16(9): 1079-1085.
[14] Bygott J. M. and Chiodini P. L. (2009). Praziquantel: neglected drug? Ineffective treatment? Or therapeutic choice in cystic hydatid disease? *Acta Trop*, 111: 95-101.
[15] Caffrey, C. R., Rohwer, A., Oellien, F., Marhöfer, R. J., Braschi, S., Oliveira, G., McKerrow, J. H. and Selzer, P. M. (2009). A comparative chemogenomics strategy to predict potential drug

targets in the metazoan pathogen, Schistosoma mansoni. *PloS one*, 4(2): e4413.

[16] Canete, R., Escobedo, A. A., Gonzalez, M. E. and Almirall, P. (2006). Randomized clinical study of five days apostrophe therapy with mebendazole compared to quinacrine in the treatment of symptomatic giardiasis in children. *World J Gastroenterol*, 12: 6366-6370.

[17] Carpio, A., Fleury, A. and Hauser, W. A. (2013). Neurocysticercosis: Five new things. *Neurol Clin Pract*, 3(2): 118-125.

[18] Chang, L., Chiang, S. H. and Saltiel, A. R. (2004). Insulin signaling and the regulation of glucose transport. *Mol Med*, 10 (7-12): 65-71.

[19] Chen, C., Jack, J. and Garofalo, R. S. (1996). The Drosophila insulin receptor is required for normal growth. *Endocrinology*, 137 (3): 846-856.

[20] Choi, G. Y., Yang, H. W., Cho, S. H., Kang, D. W., Go, H., Lee, W. C., Lee, Y. J., Jung, S. H., Kim, A. N. and Cha, S. W. (2008). Acute drug-induced hepatitis caused by albendazole. *J Korean Med Sci*, 23(5): 903-905.

[21] Cobo, F., Yarnoz, C., Sesma, B., Fraile, P., Aizcorbe, M., Trujillo, R., Diaz-de-Liano, A. and Ciga, M. A. (1998). Albendazole plus praziquantel versus albendazole alone as a pre-operative treatment in intra-abdominal hydatisosis caused by *Echinococcus granulosus*. *Trop Med Int Health*, 3: 462-466.

[22] Commission on Tropical Diseases of the International League Against Epilepsy. (1994). Relationship between epilepsy and tropical diseases. *Epilepsia*, 35: 89-93.

[23] Corona, T., Lugo, R., Medina, R. and Sotelo, J. (1996). Single-day praziquantel therapy for neurocysticercosis. *N Engl J Med*, 334: 125.

[24] Cruz, I., Cruz, M. E., Carrasco, F. and Horton, J. (1995). Neurocysticercosis: optimal dose treatment with albendazole. *J Neurol Sci*, 133(1-2):152-154.

[25] Cuello-García, C. A., Roldán-Benítez, Y. M., Pérez-Gaxiola, G. and Villarreal-Careaga J. (2013). Corticosteroids for neurocysticercosis:

a systematic review and meta-analysis of randomized controlled trials. *Int J Infect Dis*, 17(8): e583-92.
[26] Davidson, R. N., den Boer, M. and Ritmeijer, K. (2009). Paromomycin. *Trans R Soc Trop Med Hyg*, 103: 653-660.
[27] Day, T. A., Haithcock, J., Kimber, M. and Maule, A. G. (2000). Functional ryanodine receptor channels in flatworm muscle fibres. Parasitology, 120 (Pt 4): 417-422.
[28] Del Brutto O. H. (1997). Albendazole therapy for subarachnoid cysticerci: clinical and neuroimaging analysis of 17 patients. *J Neurol Neurosurg Psychiatry*, 62(6): 659-661.
[29] Del Brutto O. H. (2012). Neurocysticercosis: a review. *Scientific World Journal*, 2012, 159821.
[30] Del Brutto, O. H. (2020). Current approaches to cysticidal drug therapy for neurocysticercosis. *Expert Rev Anti Infect Ther*, 18(8): 789-798.
[31] Del Brutto, O. H. and Garcia, H. H. (2014). Control and perspectives for elimination of *Taenia solium* taeniosis/cysticercosis. In: Del Brutto OH, Garcia HH, editors. *Cysticercosis of the human nervous system*. Berlin: Springer, pp. 125-135.
[32] Del Brutto, O. H., Campos, X., Sánchez, J. and Mosquera, A. (1999). Single-day praziquantel versus 1-week albendazole for neurocysticercosis. *Neurology*, 52(5): 1079-1081.
[33] Del Brutto, O. H., Sotelo, J. and Román, G. C (1998). *Neurocysticercosis: A Clinical Handbook*. Lisse, The Netherlands: Swets & Zeitlinger.
[34] Del Brutto, O. H., Sotelo, J. and Roman, G. C. (1993). Therapy for neurocysticercosis: a reappraisal. *Clin Infect Dis*, 17(4): 730-735.
[35] Delgado-García, G., Méndez-Zurita, V. A., Bayliss, L., Flores-Rivera, J. and Fleury, A. (2019). Neurocysticercosis: mimics and chameleons. Practical neurology, 19(2): 88-95.
[36] den Hollander, P., Savage, M. I. and Brown, P. H. (2013). Targeted therapy for breast cancer prevention. *Front Oncol*, 3:250.
[37] Escobar, A. (1983). The pathology of neurocysticercosis. In: Palacios E., Rodriguez-Carbajal K. J., Taveras J. (eds), *Cysticercosis*

of the central nervous system, Charles C. Thomas: Springfield, IL, 27-54.

[38] Escobedo, G., Larralde, C., Chavarria, A., Cerbón, M. A., and Morales-Montor, J. (2004). Molecular mechanisms involved in the differential effects of sex steroids on the reproduction and infectivity of Taenia crassiceps. *J Parasitol*, 90: 1235-1244.

[39] Escobedo, G., Romano, M. C. and Morales-Montor, J. (2009). Differential in vitro effects of insulin on *Taenia crassiceps* and *Taenia solium cysticerci*. *Journal of helminthology*, 83(4), 403–412.

[40] Escobedo, G., Camacho-Arroyo, I., Hernández-Hernández, O. T., Ostoa-Saloma, P., García-Varela, M. and Morales-Montor, J. (2010a). Progesterone induces scolex evagination of the human parasite *Taenia solium*: evolutionary implications to the host-parasite relationship. *J Biomed Biotechnol*, 2010: 591079.

[41] Escobedo, G., Soldevila, G., Ortega-Pierres, G., Chávez-Ríos, J. R., Nava, K., Fonseca-Liñán, R., López-Griego, L., Hallal-Calleros, C., Ostoa-Saloma, P. and Morales-Montor, J. (2010b). A new MAP kinase protein involved in estradiol-stimulated reproduction of the helminth parasite *Taenia crassiceps*. *J Biomed Biotechnol*, 2010: 747121.

[42] Escobedo, G., Palacios-Arreola, M. I., Olivos, A., López-Griego, L. and Morales-Montor, J. (2013). Tamoxifen treatment in hamsters induces protection during taeniosis by *Taenia solium*. *Biomed Res Int*, 2013: 280496.

[43] Esquivel-Velazquez, M., Ostoa-Saloma, P., Morales-Montor, J., Hernandez-Bello, R. and Larralde, C. (2011). Immunodiagnosis of neurocysticercosis: ways to focus on the challenge. *J Biomed Biotechnol*, 2011: 516042.

[44] Fogang, Y. F., Savadogo, A. A., Camara, M., Toffa, D. H., Basse, A., Sow, A. D. and Ndiaye, M. M. (2015). Managing neurocysticercosis: challenges and solutions. *Int J Gen Med*, 8: 333-344.

[45] Frackowiak, M., Sharma, M., Singh, T., Mathew, A. and Michael, B. D. (2019). Antiepileptic drugs for seizure control in people with

neurocysticercosis. *Cochrane Database Syst Rev*, 10(10): CD009027.
[46] Garcia, H. H. and Del Brutto, O. H. (2003). Imaging findings in neurocysticercosis. *Acta Trop*, 87(1): 71-78.
[47] García, H. H., Evans, C. A., Nash, T. E., Takayanagui, O. M., White, A. C., Jr, Botero, D., Rajshekhar, V., Tsang, V. C., Schantz, P. M., Allan, J. C., Flisser, A., Correa, D., Sarti, E., Friedland, J. S., Martinez, S. M., Gonzalez, A. E., Gilman, R. H. and Del Brutto, O. H. (2002). Current consensus guidelines for treatment of neurocysticercosis. *Clin Microbiol Rev*, 15(4): 747–756.
[48] Garcia, H. H., Gilman, R. H., Horton, J., Martinez, M., Herrera, G., Altamirano, J., Cuba, J. M., Rios-Saavedra, N., Verastegui, M., Boero, J. and Gonzalez, A. E. (1997). Albendazole therapy for neurocysticercosis: a prospective double-blind trial comparing 7 versus 14 days of treatment. Cysticercosis Working Group in Peru. *Neurology*, 48: 1421-1427.
[49] Garcia, H. H., Lescano, A. G., Gonzales, I., Bustos, J. A., Pretell, E. J., Horton, J., Saavedra, H., Gonzalez, A. E., Gilman, R. H. and Cysticercosis Working Group in Peru (2016). Cysticidal Efficacy of Combined Treatment with Praziquantel and Albendazole for Parenchymal Brain Cysticercosis. *Clin infect dis: an official publication of the Infectious Diseases Society of America*, 62(11): 1375–1379.
[50] Garcia, H. H., Lescano, A. G., Lanchote, V. L., Pretell, E. J., Gonzales, I., Bustos, J. A., Takayanagui, O. M., Bonato, P. S., Horton, J., Saavedra, H., Gonzalez, A. E., Gilman, R. H., and Cysticercosis Working Group in Peru. (2011). Pharmacokinetics of combined treatment with praziquantel and albendazole in neurocysticercosis. *Br J Clin Pharmacol*, 72(1): 77-84.
[51] Hemer, S., et al., 2014. Host insulin stimulates *Echinococcus multilocularis* insulin signaling pathways and larval development. *BMC. Biol.* 12 (5).
[52] Hoberg, E. P. (2002). *Taenia* tapeworms: their biology, evolution and socioeconomic significance. *Microbes Infect*, 4: 859-866.

[53] Holmstedt, B., Nordgren, I., Sandoz, M., & Sundwall, A. (1978). Metrifonate. Summary of toxicological and pharmacological information available. *Arch Toxicol*, 41: 3-29.

[54] Hu, Y., Xiao, S. H. and Aroian, R. V. (2009). The new anthelmintic tribendimidine is an L-type (levamisole and pyrantel) nicotinic acetylcholine receptor agonist. *PLoS Negl Trop Dis*, 3(8): e499.

[55] Ito, A., Yanagida, T. and Nakao, M. (2016). Recent advances and perspectives in molecular epidemiology of *Taenia solium* cysticercosis. *Infect Genet Evol*, 40: 357-367.

[56] Jung, H., M. Hurtado, M. Sanchez, M. T. Medina, and J. Sotelo. (1990). Plasma and cerebrospinal fluid levels of albendazole and praziquantel in patients with neurocysticercosis. *Clin Neuropharmacol*, 13: 559-564.

[57] Jung, H., M. Hurtado, M. T. Medina, M. Sanchez, and J. Sotelo. (1990). Dexamethasone increases plasma levels of albendazole. *J Neurol*, 237: 279-280.

[58] Khayath, N., Vicogne, J., Ahier, A., BenYounes, A., Konrad, C., Trolet, J., Viscogliosi, E., Brehm, K. and Dissous, C. (2007). Diversification of the insulin receptor family in the helminth parasite *Schistosoma mansoni*. *FEBS J*, 274(3): 659-676.

[59] Kohn, A. B., Anderson, P. A., Roberts-Misterly, J. M. and Greenberg, R. M. (2001). *Schistosome calcium* channel beta subunits. Unusual modulatory effects and potential role in the action of the antischistosomal drug praziquantel. *J Biol Chem*, 276(40): 36873-36876.

[60] Kohn, A. D., Summers, S. A., Birnbaum, M. J. and Roth, R. A. (1996). Expression of a constitutively active Akt Ser/Thr kinase in 3T3-L1 adipocytes stimulates glucose uptake and glucose transporter 4 translocation. *J. Biol. Chem*, 271 (49): 31372-31378.

[61] Konrad, C., Kroner, A., Spiliotis, M., Zavala-Góngora, R., & Brehm, K. (2003). Identification and molecular characterisation of a gene encoding a member of the insulin receptor family in *Echinococcus multilocularis*. *Int J Parasitol*, 33 (3): 301–312.

[62] Koul, P. A., Wahid, A., Bhat, M. H., Wani, J. I. and Sofi, B. A. (2000). Mepacrine therapy in niclosamide resistant taeniasis. *J Assoc Physicians India*, 48(4): 402-403.
[63] Kramer, L. D. (1995). Medical treatment of cysticercosis: ineffective. *Arch Neurol*, 52: 101-112.
[64] Kulke, D., Krucken, J., Welz, C., von Samson-Himmelstjerna, G. and Harder, A. (2012). In vivo efficacy of the anthelmintic tribendimidine against the cestode *Hymenolepis microstoma* in a controlled laboratory trial. *Acta Trop*, 123: 78-84.
[65] Lanchote, V. L., Garcia, F. S., Dreossi, S. A. and Takayanagui, O. M. (2002). Pharmacokinetic interaction between albendazole sulfoxide enantiomers and antiepileptic drugs in patients with neurocysticercosis. *Ther Drug Monit*, 24(3): 338-345.
[66] Lateef, M., Zargar, S. A., Khan, A. R., Nazir, M. and Shoukat, A. (2008). Successful treatment of niclosamide- and praziquantel-resistant beef tapeworm infection with nitazoxanide. *Int J Infect Dis*, 12(1): 80-82.
[67] Li, A. H., Moon, S. U., Park, Y. K., Na, B. K., Hwang, M. G., Oh, C. M., Cho, S. H., Kong, Y., Kim, T. S. and Chung, P. R. (2006). Identification and characterization of a cathepsin L-like cysteine protease from *Taenia solium* metacestode. *Vet parasitol*, 141(3-4): 251-259.
[68] Lightowlers, M. W. (2013). Control of Taenia solium taeniasis/ cysticercosis: past practices and new possibilities. *Parasito-logy*, 140: 1566-1577.
[69] López-Gómez, M., Castro, N., Jung, H., Sotelo, J. and Corona, T. (2001). Optimization of the single-day praziquantel therapy for neurocysticercosis. *Neurology*, 57(10): 1929-1930.
[70] Marks, N. J. and Maule, A. G. (2010). Neuropeptides in helminths: occurrence and distribution. *Adv Exp Med Biol*, 692: 49-77.
[71] Matthaiou, D. K., Panos, G., Adamidi, E. S. and Falagas, M. E. (2008). Albendazole versus praziquantel in the treatment of neurocysticercosis: a meta-analysis of comparative trials. *PLoS Negl Trop Dis*, 2(3): e194. Pushpakom S., Iorio F., Eyers P. A., Escott K.

J., Hopper S., Wells A., Doig A., Guilliams T., Latimer J., McNamee C., et al. Drug repurposing: Progress, challenges and recommendations. *Nat. Rev. Drug Discov.* 2019; 18:41–58. doi: 10.1038/nrd.2018.168.

[72] McVeigh, P., McCammick, E., McCusker, P., Wells, D., Hodgkinson, J., Paterson, S., Mousley, A., Marks, N. J., & Maule, A. G. (2018). Profiling G protein-coupled receptors of Fasciola hepatica identifies orphan rhodopsins unique to phylum Platyhelminthes. *Int J Parasitol Drugs Drug Resist*, 8(1): 87-103.

[73] Nash, T. E. (2003). Human case management and treatment of cysticercosis. *Acta Trop*, 87: 61–69.

[74] Nash, T. E. and Garcia, H. H. (2011). Diagnosis and treatment of neurocysticercosis. *Nat Rev Neurol*, 7(10): 584-594.

[75] Nawaratna, S., You, H., Jones, M. K., McManus, D. P. and Gobert, G. N. (2018). Calcium and Ca2+/Calmodulin-dependent kinase II as targets for helminth parasite control. *Biochem Soc trans*, 46(6): 1743-1751.

[76] O'Neal, S. E. and Flecker, R. H. (2015). Hospitalization frequency and charges for neurocysticercosis, United States, 2003–2012. *Emerg Infect Dis*, 21: 969-976.

[77] Oldham, S. and Hafen, E. (2003). Insulin/IGF and target of rapamycin signaling: a TOR de force in growth control. *Trends Cell Biol*, 13 (2): 79-85.

[78] Opatrny, L., Prichard, R., Snell, L. and Maclean, J. D. (2005). Death related to albendazole-induced pancytopenia: case report and review. *Am J Trop Med Hyg*, 72: 291-294.

[79] Overbosh, D., van de Nes, J. C., Groll, E., Diekmann, H. W., Polderman, A. M. and Mattie, H. (1987). Penetration of praziquantel into cerebrospinal fluid and cysticerci in human cysticercosis. *Eur J Clin Pharmacol*, 33: 287-292.

[80] Palomares-Alonso, F., Piliado, J. C., Palencia, G., Ortiz-Plata, A. and Jung-Cook, H. (2007). Efficacy of nitazoxanide, tizoxanide and tizoxanide/albendazole sulphoxide combination against *Taenia crassiceps* cysts. *J Antimicrob Chemother*, 59(2): 212-218.

[81] Panic, G., Duthaler, U., Speich, B. and Keiser, J. (2014). Repurposing drugs for the treatment and control of helminth infections. *Int J Parasitol Drugs Drug Resist*, 4(3): 185-200.

[82] Prodjinotho, U. F., Lema, J., Lacorcia, M., Schmidt, V., Vejzagic, N., Sikasunge, C., Ngowi, B., Winkler, A. S., and Prazeres da Costa, C. (2020). Host immune responses during Taenia solium Neurocysticercosis infection and treatment. *PLoS Negl Trop Dis*, 14(4): e0008005.

[83] Prole, D. L. and Taylor, C. W. (2011). Identification of intracellular and plasma membrane calcium channel homologues in pathogenic parasites. *PLoS ONE*, 6: e26218.

[84] Pushpakom, S., Iorio, F., Eyers, P. A., Escott, K. J., Hopper, S., Wells, A., Doig, A., Guilliams, T., Latimer, J., McNamee C., Norris, A., Sanseau, P., Cavalla, D. and Pirmohamed, M. (2019). Drug repurposing: Progress, challenges and recommendations. *Nat Rev Drug Discov*, 18: 41-58.

[85] Rajshekhar, V. (2010). Surgical management of neurocysticercosis. *Int J Surg*, 8(2): 100-104.

[86] Rodriguez, S., Wilkins, P. and Dorny, P. (2012). Immunological and molecular diagnosis of cysticercosis. *Pathog Glob Health*, 106(5): 286-298.

[87] Romo, M. L., Carpio, A. and Kelvin, E. A. (2014). Routine drug and food interactions during antihelminthic treatment of neurocysticercosis: a reason for the variable efficacy of albendazole and praziquantel? *J Clin Pharmacol*, 54(4): 361-367.

[88] Rossignol, J. F. (1981). Albendazole: estudios clínicos realizados en Francia y Africa Occidental. Informe sobre 1034 casos. *Compendium de Investigaciones Clínicas Latinoamericanas*, 1:117-125. [Albendazole: clinical studies conducted in France and West Africa. Report on 1034 cases. *Latin American Clinical Research Compendium*]

[89] Rossignol, J. F. and Maisonneuve, H. (1984). Albendazole: a new concept in the control of intestinal helminthiasis. *Gastroenterol Clin Biol*, 8(6-7): 569-76.

[90] Rueda, A., Sifuentes, C., Gilman, R. H., Gutiérrez, A. H., Piña, R., Chile, N., Carrasco, S., Larson, S., Mayta, H., Verástegui, M., Rodriguez, S., Gutiérrez-Correa, M., García, H. H., Sheen, P. and Zimic, M. (2011). TsAg5, a *Taenia solium* cysticercus protein with a marginal trypsin-like activity in the diagnosis of human neurocysticercosis. *Mol biochem parasitol*, 180(2): 115-119.

[91] Salem, H. H. and el-Allaf, G. (1969). Treatment of *Taenia saginata* and Hymenolepis nana infections with paromomycin. *Trans R Soc Trop Med Hyg*, 63: 833-836.

[92] Santos, R., Ursu, O., Gaulton, A., Bento, A. P., Donadi, R. S., Bologa, C. G., Karlsson, A., Al-Lazikani, B., Hersey, A., Oprea, T. I. and Overington, J. P. (2017). A comprehensive map of molecular drug targets. *Nat Rev Drug Discov*, 16 19-34.

[93] Shareef, P. A. and Abidi, S. M. (2014). Cysteine protease is a major component in the excretory/secretory products of *Euclinostomum heterostomum* (Digenea: Clinostomidae). *Parasitol Res*, 113: 65-71.

[94] Sharma, M., Singh, T. and Mathew, A. (2015). Antiepileptic drugs for seizure control in people with neurocysticercosis. *Cochrane Database Syst Rev*, 10: CD009027.

[95] Singh, G. and Sander, J. (2004). Anticysticercal treatment and seizures in neurocysticercosis. *Lancet Neurol*, 3: 207-208.

[96] Singh, N. and Narayan, S. (2011). Nitazoxanide: A Broad Spectrum Antimicrobial. *Med J Armed Forces India*, 67(1): 67-68.

[97] Singhi, P. Neurocysticercosis. (2011). *Ther Adv Neurol Disord*, 4(2): 67-81.

[98] Sinha, S. and Sharma, B. S. (2009). Neurocysticercosis: a review of current status and management. *J Clin Neurosci*, 16: 867-876.

[99] Smith, J. L. *Taenia solium* neurocysticercosis. (1994). *J Food Prot*, 57: 831–844.

[100] Sotelo, J. and Jung, H. (1998). Pharmacokinetic optimisation of the treatment of neurocysticercosis. *Clin Pharmacokinet*, 34(6): 503-515.

[101] Sotelo, J., del Brutto, O. H., Penagos, P., Escobedo, F., Torres, B., Rodriguez-Carbajal, J. and Rubio-Donnadieu, F. (1990). Comparison

of therapeutic regimen of anticysticercal drugs for parenchymal brain cysticercosis. *J Neurol*, 237(2): 69-72.

[102] Sotelo, J., Escobedo, F., Rodriguez-Carbajal, J., Torres, B. and Rubio-Donnadieu, F. (1984). Therapy of parenchymal brain cysticercosis with praziquantel. *N Engl J Med*, 310: 1001-1007.

[103] Sotelo, J., Escobedo, F. and Penagos, P. (1988a). Albendazole vs. praziquantel for therapy for neurocysticercosis. A controlled trial. *Arch Neurol*, 45: 532-534.

[104] Sotelo, J., Penagos, P., Escobedo, F. and Del Brutto, O. H. (1988b). Short course of albendazole therapy for neurocysticercosis. *Arch Neurol*, 45: 1130-1133.

[105] Sotelo, J., Torres, B., Rubio-Donnadieu, F., Escobedo, F., and Rodriguez-Carbajal, J. (1985). Praziquantel in the treatment of neurocysticercosis: long-term follow-up. *Neurology*, 35(5): 752-755.

[106] Srikanth, M. and Kessler, J. A. (2012). Nanotechnology-novel therapeutics for CNS disorders. *Natur Rev Neurol*, 8(6), 307–318.

[107] Steinmann, P., Zhou, X. N., Du, Z. W., Jiang, J. Y., Xiao, S. H., Wu, Z. X., Zhou, H. and Utzinger, J. (2008). Tribendimidine and Albendazole for Treating Soil-Transmitted Helminths, *Strongyloides stercoralis* and *Taenia* spp.: Open-Label Randomized Trial. PLoS Negl Trop Dis, 2: e322.

[108] Stettler, M., Fink, R., Walker, M., Gottstein, B., Geary, T. G., Rossignol, J. F. and Hemphill, A. (2003). In vitro parasiticidal effect of Nitazoxanide against *Echinococcus multilocularis* metacestodes. *Antimicrob Agents Chemother*, 47: 467-474.

[109] Swulius, M. T. and Waxham, M. N. (2008). Ca(2+)/calmodulin-dependent protein kinases. *Cell Mol Life Sci*, 65: 2637-2657.

[110] Symeonidou, I., Arsenopoulos, K., Tzilves, D., Soba, B., Gabriël, S. and Papadopoulos, E. (2018). Human taeniasis/cysticercosis: a potentially emerging parasitic disease in Europe. *Ann Gastroenterol*, 31(4): 406-412.

[111] Takayanagui, O. M. and Jardim, E. (1992). Therapy for neurocysticercosis. Comparison between albendazole and praziquantel. *Arch Neurol*, 49(3): 290-294.

[112] Takayanagui, O. M., Odashima, N. S., Bonato, P. S., Lima, J. E. and Lanchote, V. L. (2011). Medical management of neurocysticercosis. *Expert Opin Pharmacother*, 12: 2845-2856.

[113] Tellez-Zenteno, J. F., AND Hernandez-Ronquillo, L. (2017). Epidemiology of neurocysticercosis and epilepsy, is everything described?. *Epilepsy & behavior: E&B*, 76, 146-150.

[114] Tsai, I. J., Zarowiecki, M. and Holroyd, N. (2013). The genomes of four tapeworm species reveal adaptations to parasitism. *Nature*, 496: 57-63.

[115] Tschen, E. A. and Smith, E. B. (1981). Cutaneous cysticercosis treated with metrifonate. *Arch Dermatol*, 117: 507-509.

[116] Ueki, K., Yamamoto-Honda, R., Kaburagi, Y., Yamauchi, T., Tobe, K., Burgering, B. M., Coffer, P. J., Komuro, I., Akanuma, Y., Yazaki, Y. and Kadowaki, T. (1998). Potential role of protein kinase B in insulin-induced glucose transport, glycogen synthesis, and protein synthesis. *J Biol Chem*, 273 (9): 5315-5322.

[117] Vaca-Paniagua, F., Torres-Rivera, A., Parra-Unda, R. and Landa, A. (2008). *Taenia solium*: antioxidant metabolism enzymes as targets for cestocidal drugs and vaccines. *Curr Top Med Chem*, 8(5): 393-399.

[118] Van Nassauw, L., Toovey, S., Van Op den Bosch, J., Timmermans, J. P. and Vercruysse, J. (2008). Schistosomicidal activity of the antimalarial drug, mefloquine, in Schistosoma mansoni-infected mice. *Travel Med Infect Dis*, 6: 253-258.

[119] Vargas-Parada, L. and Laclette, J. P. (1999). Role of the calcareous corpuscles in cestode physiology: a review. *Rev Latinoam Microbiol*, 41(4): 303-307.

[120] Vargas-Villavicencio, J. A., Larralde, C. and Morales-Montor, J. (2008). Treatment with dehydroepiandrosterone in vivo and in vitro inhibits reproduction, growth and viability of *Taenia crassiceps* metacestodes. *Int J Parasitol*, 38: 775-781.

[121] Vazquez, M. L., H. Jung, and J. Sotelo. (1987). Plasma levels of praziquantel decrease when dexamethasone is given simultaneously. *Neurology*, 37: 1561-1562.

[122] Vermund, H. S., MacLeod, S. and Goldstein, R. G. (1986). Taeniasis unresponsive to a single dose of niclosamide: case report of persistent infection with *Taenia saginata* and a review of therapy. *Rev Infect Dis*, 8: 423-426.
[123] Vielma, J. R., Urdaneta-Romero, H., Villarreal, J. C., Paz, L. A., Gutiérrez, L. V., Mora, M. and Chacín-Bonilla L. (2014) Neurocysticercosis: Clinical Aspects, Immunopathology, Diagnosis, Treatment and Vaccine Development. *Epidemiol*, 4: 156.
[124] Walker, M., Rossignol, J. F., Torgerson, P. and & Hemphill, A. (2004). In vitro effects of nitazoxanide on *Echinococcus granulosus* protoscoleces and metacestodes. *J Antimicrob Chemother*, 54: 609-616.
[125] Wang, L., Liang, P., Wei, Y., Zhang, S., Guo, A., Liu, G., Ehsan, M., Liu, M. and Luo, X. (2020). *Taenia solium* insulin receptors: promising candidates for cysticercosis treatment and prevention. *Acta Trop*, 209: 105552.
[126] White, A. C. (2000). Neurocysticercosis: updates on epidemiology, pathogenesis, diagnosis, and management. *Annu Rev Med*, 51: 187-206.
[127] White, A. C., Jr, Coyle, C. M., Rajshekhar, V., Singh, G., Hauser, W. A., Mohanty, A., Garcia, H. H. and Nash, T. E. (2018). Diagnosis and Treatment of Neurocysticercosis: 2017 Clinical Practice Guidelines by the Infectious Diseases Society of America (IDSA) and the American Society of Tropical Medicine and Hygiene (ASTMH). *Clin Infect Dis*, 66(8): 1159-1163.
[128] Willms, K., Robert, L. and Caro, J. A. (2003). Ultrastructure of smooth muscle, gap junctions and glycogen distribution in *Taenia solium* tapeworms from experimentally infected hamsters. *Parasitol Res*, 89(4): 308-316.
[129] Winkler, A. S. (2012). Neurocysticercosis in sub-Saharan Africa: a review of prevalence, clinical characteristics, diagnosis, and management. *Pathog Glob Health*, 106: 261-274.
[130] Winkler, A. S. (2013). Epilepsy and neurocysticercosis in sub-Saharan Africa. In: Foyaca-Sibat H, editor. *Novel Aspects on*

Cysticercosis and Neurocysticercosis. Rijeka, Croatia: InTech, 307-340.

[131] Xiao, S. H., Hui-Ming, W., Tanner, M., Utzinger, J. and Chong, W. (2005). Tribendimidine: A promising, safe and broad-spectrum anthelmintic agent from China. *Acta Trop*, 94: 1-14.

[132] Xiao, S. H., Jian, X., Tanner, M., Yong-Nian, Z., Keiser, J., Utzinger, J. and Hui-Qiang, Q. (2008). Artemether, artesunate, praziquantel and tribendimidine administered singly at different dosages against *Clonorchis sinensis*: a comparative *in vivo* study. *Acta Trop*, 106: 54-59.

[133] Yan H. B., Lou Z. Z., Li L. et al. (2014) Yan, H. B., Lou, Z. Z., Li, L., Brindley, P. J., Zheng, Y., Luo, X., Hou, J., Guo, A., Jia, W. Z. and Cai, X. (2014). Genome-wide analysis of regulatory proteases sequences identified through bioinformatics data mining in *Taenia solium*. *BMC genomics*, 15: 428.

[134] You, H., Zhang, W., Jones, M. K., Gobert, G. N., Mulvenna, J., Rees, G., Spanevello, M., Blair, D., Duke, M., Brehm, K. and McManus, D. P. (2010). Cloning and characterisation of *Schistosoma japonicum* insulin receptors. *PloS one*, 5(3):e9868.

[135] Zimic, M., Pajuelo, M., Rueda, D., López, C., Arana, Y., Castillo, Y., Calderón, M., Rodriguez, S., Sheen, P., Vinetz, J. M., Gonzales, A., García, H. H., Gilman, R. H. and Cysticercosis Working Group in Perú (2009). Utility of a protein fraction with cathepsin L-Like activity purified from cysticercus fluid of *Taenia solium* in the diagnosis of human cysticercosis. *Am J Trop Med Hyg*, 80: 964-970.

In: Neurocysticercosis
Editor: Mark A. Chavez

ISBN: 978-1-53619-791-4
© 2021 Nova Science Publishers, Inc.

Chapter 2

THE CHALLENGE OF DECIPHERING CERTAINTY FROM AMBIGUITY FOR THE LABORATORY DIAGNOSIS OF NEUROCYSTICERCOSIS

Rimanpreet Kaur, Naina Arora*, Suraj Singh Rawat, Anand Kumar Keshri, Neha Singh, Avinash Singh, Shweta Tripathi and Amit Prasad†*
School of Basic Sciences, Indian Institute of Technology Mandi, Mandi, Himachal Pradesh, India

1. INTRODUCTION

Neurocysticercosis (NCC) is a major cause of acquired epilepsy in developing regions of the world where socio-economic indexes are low

---

* These author contributed equally.
† Corresponding Author's E-mail: amitprasad@iitmandi.ac.in.

and hygiene and health awareness is less among the communities. According to a WHO report, NCC causes approximately 50,000 human deaths per year, and in 2014 the parasite *T. solium* was ranked first on global scale of food born parasites. The WHO Food Borne Disease Burden Epidemiology Reference Group 2015 had identified "*T. solium* as a leading cause of deaths by food borne diseases considerable to 2.8 million disability adjusted life years loss (DALYs)." Another recent study had estimated that NCC accounts for at least 5% of all avoidable epilepsy cases (WHO 2019) across globe and approximately 30% of new acquired epilepsy cases in endemic areas (Prasad et al. 2008; Prasad et al. 2009). In a WHO report of 2015, approximately 0.317%, 0.076% and 0.597% (highest) of total population is estimated to be at risk of infection in Peru, Ecuador and India, respectively (Ito et al. 2003; Torgerson et al. 2015). Hence, for endemic areas it is a major cause of active epilepsy (Watts et al. 2014; Torgerson et al. 2015; Prasad et al. 2009). In Global Burden of Disease Study (GBDS) 2010 report, it was estimated that 0.07 DALYs are lost per 1000 people globally due to occurrence of NCC, but still this data was considered to be an under estimation due to complexity associated with the accurate diagnosis of NCC.

The diagnosis of NCC is complex due to its varied clinical presentation, which ranges from mild headache to severe recurring seizure; and in some extreme rare cases mortality is also reported (Ndimubanzi et al. 2010; Arora et al. 2019). The clinical manifestations of NCC varies from person to person due to differences in the load and stage of the loaded parasite in the person. This extreme variation in clinical outcome of infection also depends upon several environmental and host genetic/immunological factors which still need to be identified accurately (Arora et al. 2018; 2019). The most significant clinical sign is occurrence of acute symptomatic multiple episodes of seizures which is observed in 80% of cases (Ndimubanzi et al. 2010); the other conditions described in symptomatic infections are headache, chronic meningitis, focal neurological deficit hydrocephalus, spinal and ocular cysts, increased intracranial pressure or cognitive decline, psychological disorders etc. (Rajshekhar et al. 2006; Oscar H. Del Brutto et al. 2017; Garcia, Nash, and

Del Brutto 2014). Considering the pleiomorphic clinical presentation and difficulty in diagnosis of NCC, a detailed guideline was made by International Working Group on NCC of International League Against Epilepsy (ILAE) in 2004 that considered clinical, neuroimaging, immunological and epidemiological data to identify the definitive or probable cases of NCC (Del Brutto et al. 2001; Montano et al. 2005). The guideline was revised again in 2017 to make it simple (O. H. Del Brutto et al. 2017), these guidelines are mentioned below.

2. GUIDELINES FOR NCC DIAGNOSIS

Table 1. Proposed diagnostic criteria for neurocysticercosis (2004)

Categories of criteria	Criteria
Absolute	1. Histological demonstration of the parasite from biopsy of a brain or spinal-cord lesion 2. Cystic lesions showing the scolex on CT or MRI 3. Direct visualization of subretinal parasites by funduscopic examination
Neuroimaging Criteria	
Major neuroimaging criteria	1. Cystic lesions without a discernible scolex 2. Enhancing lesions 3. Multilobulated cystic lesions in the subarachnoid space 4. Typical parenchymal brain calcifications
Confirmative neuroimaging criteria	1. Resolution of cystic lesions after cysticidal drug therapy 2. Spontaneous resolution of single small enhancing lesions 3. Migration of ventricular cysts documented on sequential neuroimaging studies
Minor Neuroimaging criteria	1. Obstructive hydrocephalus (symmetric or asymmetric) or abnormal enhancement of basal leptomeninges
Clinical/exposure criteria	
Major clinical/exposure	2. Detection of specific anticysticercal antibodies or cysticercal antigens by well-standardized immunodiagnostic tests 3. Cysticercosis outside the central nervous system 4. Evidence of a household contact with *T. solium* infection
Minor clinical/exposure	1. Clinical manifestations suggestive of neurocysticercosis 2. Individuals coming from or living in an area where cysticercosis is endemic

Diagnosis of NCC is severely compromised by its variable clinical presentations. The ILAE has given a broad diagnostic criterion to assist clinicians for the accurate diagnosis of NCC (Table 1). This diagnostic criterion has absolute, neuroimaging and clinical or exposure criteria, where the combination of evidences in these criteria is used to classify patient as NCC positive or negative. Accurate interpretation of these criteria authorizes two degrees of diagnostic certainty, definitive or probable. Absolute diagnosis of NCC requires direct visualization of the parasite in tissues, either by indicating parasite in brain tissue by histology or by demonstration of scolex in cystic lesions using neuroimaging tools (MRI/CT). Whereas the initial set of diagnostic criteria that was proposed had absolute, major, minor and epidemiological criteria.

The above guidelines have been recently revised to increase the degree of certainty of diagnosis (Del Brutto et al. 2017) and had been given in Table 2.

Table 2. Revised degree of certainty for the diagnosis of neurocysticercosis (2017)

Diagnostic certainty	Criteria
Definitive	1. One absolute criterion.
	2. Two major neuroimaging criteria plus any clinical/exposure criteria.
	3. One major and one confirmative neuroimaging criteria plus any clinical/exposure criteria.
	4. One major neuroimaging criteria plus two clinical/exposure criteria (including at least one major clinical/exposure criterion), together with the exclusion of other pathologies producing similar neuroimaging findings
Probable	1. Presence of one major neuroimaging criteria plus any two clinical/exposure criteria
	2. Presence of one minor neuroimaging criteria plus at least one major clinical/exposure criteria.

3. AVAILABLE DIAGNOSTIC TOOLS FOR NCC

In the revised diagnostic criteria too, as was the case in the previous guideline, the direct visualization of cystic lesion through MRI or CT scan along with the presence of antibodies in patient serum in clinical criteria is considered as absolute and confirmatory. The preference of neuroimaging techniques depends on the location and stage of the cystic lesion in the CNS such as Fluid-attenuated inversion recovery (FLAIR) MRI is better to visualize the cyst located at extra ventricular subarachnoid space and cisternal lesions on the other hand CT scan is better to visualize calcified cysts in the brain (Frigon et al. 2004).

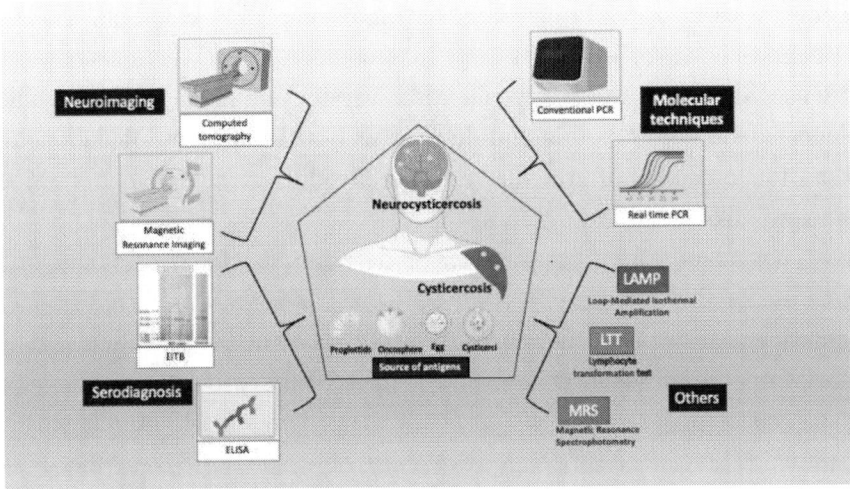

Figure 1. Conventional and unconventional methods for the detection of NCC infection

3.1. Neuroimaging Techniques (MRI and CT)

Application of neuroimaging modalities such as computed tomography (CT) and magnetic resonance imaging (MRI) have enhanced the accuracy of the NCC diagnosis. The neuroimaging diagnostic findings for NCC depends on the stage of larval development and involution, type of cysticercus and, number and location of cysts in the brain. By applying neuroimaging tools, four stages of cyst formation have been defined and the imaging findings in each case suggest fundamental changes in the host response and disease process. CT defines vesicular stage of the cyst as a lesion with hypo-dense containing a hyper-intense small scolex along with non-enhancing or meekly enhancing cyst wall. At the colloidal vesicular stage, the parasite's larva begins to disintegrate and the host's immune reactivity creates a fibrous capsule to form the surrounding parenchymal edema. This stage is depicted as ring enhancing cystic lesions with hyperintense fluid content and surrounding edema on CT examination. In the granular-nodular stage, cyst further withdraws and forms a thick granulomatous nodule that appears as an enhancing nodule with mild surrounding edema. In the final calcified stage of cyst, the granulomatous lesion appears to be further shrunken and completely calcified. The *C. racemosus* is described by its large size compared to *C. cellulosae*, with absence of scolex and a protean form varying from that of a sizeable bladder delimited by a thin and delicate wall and with few lobulations to that of a complex of irregularly sized bladders arranged in a fashion appearing like a cluster of grapes.

CT is majorly used for cystic lesion detection due to cost friendly with wide availability (Gasparetto et al. 2016). CT is always preferred in poor countries (African countries) because of its capability of detecting cyst in most forms with superior specificity and sensitivity if contrast is not a major concern (Kuehnast et al. 2020). Together with EITB diagnostic method, CT scan confirms the detection of cystic lesions in the brain (Debacq et al. 2017). It has been claimed that by use of CT a sensitivity and specificity of about 95% for the diagnosis of NCC can be achieved; however its sensitivity falls much lower for ventricular or cisternal forms of the disease. In a swine study it has been reported that among the 11 infected animals CT could detect cysticeri with only 66.5% efficiency (Gonzalez et al. 1987). That further confirms that all cysticerci are not visible on CT because of its low resolution, and two or more tightly closed placed parasites may appear as one hypodense area. Viable cysts give weak signal in CT imaging as compare to degenerating and colloidal stage cysts, at degenerating/colloidal stages contrast material gives enhanced ring formation around the cystic lesions in CT images. Subarachnoid and ventricular cysts appear in CT images as well demarcated hypointense area on T2-weighted image, cysts in these location cause asymmetric or obstructive hydrocephalus which is most common finding of CT scan in

NCC (Madrazo et al. 1983; Kimura-Hayama et al. 2010b). Degenerating cyst lesion which is primarily associated for seizure occurrence in NCC.

The another neuro-imaging tool, MRI is the state of the art imaging technology and has taken the centre stage as the major technique in the routine diagnosis of many neurological and other disease processes. It has particular leads in that it is non-invasive, it does not need any ionizing radiation and has better soft-tissue resolution compared to CT. It is capable of providing both functional and morphological information. The MR image generated is based on multiple parameters and any of these can modify tissue contrast. It provides more contrasting images than CT scan to differentiate between CSF and cystic lesions, however misses in detecting calcified lesions. It is considered as the best neuro-imaging tool for the detection of viable and degenerating cysticerci, while CT is considered best for detection of calcified lesions (Garcia and Del Brutto 2013). The extra benefit of MRI application is that it can differentiate the stages of the cyst where CT disappoints. However, it is very expensive to operate so not generally available in the endemic regions hence, less preferred than CT scan (Kimura-Hayama et al. 2010c).

3.1.1. Sero-Diagnostic Tools

Serological methods are used to determine the concentration of antibody in patient serum against particular infection using specific antigens. In NCC infection, antibodies are generated against cystic lesions, earlier, whole cysticerci homogenate was used as antigens to detect antibodies in the patient serum, but it gave false positive results, because it showed cross-reactivity with other helminth antigens. Later on, with the advancement of technology, proteins were purified based on their size and isoelectric point from the whole homogenate to increase the sensitivity of the diagnostic tools.

3.1.2. Enzyme Electro-Immune Transfer Blot

EITB is an antibody based diagnostic technique, it has been previously used to diagnose schistosomiasis infection in humans and for NCC it was first described by Grogl and colleagues where they characterized 37 proteins from *T. solium* metacestode (Grogl et al. 1985). The 37 polypeptides of *T. solium* larvae showed reactivity with NCC patient serum and 31 out of 37 were documented as specific to NCC because 6 polypeptides reacted with control serum too. Out of these 31 proteins, 10 were reported as glycoproteins with different molecular weight (range from 200kDa to 16kDa) and used for diagnosis of NCC (Tsang, Brand, and Boyer 1989). Glycoproteins were first isolated using lentil-lectin, affinity-based purification method then separated via sodium dodecyl sulphate gel electrophoresis followed by transfer on nitro cellulose membrane and detection by patient serum samples. By this EITB, they identified seven major diagnostic bands of 13 kDa, 14 kDa, 18 kDa, 21 kDa, 24 kDa, 39-42 kDa and 50 kDa, that was detected by NCC patients with 98% sensitivity and 100% specificity (Tsang, Brand, and Boyer 1989). The role of glycan moieties on lower kDa glycoproteins particularly of 12,16,18, 24 and 28 kDa were analysed by deglycosylation and later it was identified that the glycan component of respective glycoproteins constitutes about 50% of the total molecular weight and can participate in the antigenicity of the glycoproteins. The study also revealed that the low molecular weight proteins can be the deglycosylated product of higher molecular weight glycoproteins (Obregón-Henao et al. 2003). We also previously reported that low molecular weight proteins are specific to NCC infection and more prominent in case of multiple cyst infection (Prasad et al. 2008).

The antibody-based tools are specific and cost-effective as compared to other tools available, however there are some limitations associated with them for diagnosing NCC accurately (Chung et al. 1999). The EITB based diagnosis is not able to effectively discriminate between active and inactive cases of NCC. Several low molecular weight protein antigens such as TsRS1, Ts18var1, Ts8B1, Ts8B2 and, Ts8B3 were reported to be useful and among all above proteins Ts8B2 protein was effective to differentiate between active and inactive cases of NCC (Feldman et al. 1990; Greene et al. 2000). In our study among the NCC patients from north India, we also found 15kDa band associated with NCC, while 17kDa was found only among degenerating state of cyst (Arora et al. 2020).

One of the most significant disadvantages of EITB is the source of LLGP antigens. The most reliable source of antigens should be the fresh cysts from the infected pig, getting that itself is a tricky task. The complex purification steps combined with lot-to-lot sample variation, it is very difficult to standardize the protocol for antigen isolation and maintain the quality of diagnostic tool (Garcia et al. 2018). The diagnostic value of glycoprotein bands can be different as originally reported depending on the source of *T. solium* antigens. All the commercially available kits use the antigens from different sources that ultimately affects the sensitivity of the test (Romo et al. 2018). *T. solium* antigens, GST-T24H and T24H-His are explored as suitable alternatives to LLGP antigens. T24H-His appeared as best suited antigens when analysed on the basis of specificity and sensitivity. The recombinant production of T24H-His protein antigen has lowered the lot to lot variation (Hernández-González et al. 2017).

Thus to conclude, EITB serves as an effective and reliable tool to diagnose NCC when neuroimaging tools are not accessible. It is less expensive and easy to perform. Though the performance of EITB is not satisfactory in case of single cyst or with calcified cyst in the brain, but use of CSF as sample may help in certain condition (Garcia et al. 2018).

3.1.3. Enzyme Linked Immunosorbent Assay (ELISA)

ELISA is the most popular method for the detection of antigen (Ag) or antibody (Ab) level in the patient's body fluid sample. It is a quantitative, sensitive and simple technique. Serum based ELISA for NCC has been very widely used due to ease of the techniques and its role in community level studies. Though this technique is simple, reliable and can be easily performed at primary health care centres, it is not considered as a diagnostic adjunct for NCC.

Still different parts of cysts like cyst wall, scolex only and cyst fluid have been used as antigen to make different kinds of ELISA by several groups with variable success (Odashima, Takayanagui and De Castro Figueiredo 2002). In a very earlier attempt, antigens prepared from either scolex or cyst wall had shown highest discriminatory potential between infected and healthy volunteers with 98% specificity and 62% sensitivity

(Espinoza et al. 1986, Fliser et al. 1975; Arambulo et al. 1978) made an ELISA from crude extracts (CE) and a partially refined fraction of antigen B (AgB) and tested serum obtained from surgically confirmed NCC patients. They found the sensitivity of this ELISA was 85% with CE and 80% with AgB while specificity was better with CE (100%) than AgB (96%). Later Coker-Vann et al. in 1984 used this technique for the sero-epidemiological studies of cysticercosis in West New Guinea patients with no false positive result. Subsequently, Ito et al. (1998) had shown that the cyst fluid (CF) had less back ground reactivity compared to CE (Ito et al. 1998). After this many ELISA's have been developed; however the major problem with these tools are that they give positive result with taeniasis cases too. Prabhakarn et al. (2004) established an ELISA from the lentin-lectin affinity purified CF and had shown that this ELISA had 80% sensitivity and 94% specificity for solitary cysticercus granuloma (SCG) (Prabhakaran et al. 2004). However none of the available tools are standardized for highly endemic communities from urban and rural population and they need to be tested for different patients base from various geological locations. Basically, for the diagnosis of NCC different types of *T. solium* Ag has been used to obtain good specificity and sensitivity, the source of antigen varied from fraction of crude cyst, cyst fluid and wall (Odashima, Takayanagui, and De Castro Figueiredo 2002) but still of no clinical importance if performed from serum.

3.1.4. Sandwich ELISA

In sandwich ELISA, Ag is immobilized on a solid surface and it captures Ab present in applied samples. Then a second enzyme conjugated Ab (reporter Ab) is used to detect the presence of captured Ab-Ag complex by adding chromogenic substrate.

The Ag-ELISA are based on the detection of circulating larval Ag in serum which reflects the severity of the infection which can be helpful to take therapeutic decision. Initially, the polyclonal antisera were raised in rabbit against crude extract of cyst which were further used in ELISA to found *T. solium* Ag in the patients CSF. This gave better result with high specificity and detected all antigenic fractions of cyst fluid present in patient's CSF. Further, monoclonal antibodies (MoAbs) were used for the detection of Ag, which considerably enhanced the performance of this assay. Harrison et al. in 1989 used a MoAb named as antiHP10 that reacted with repetitive epitope of excretory secretory glycoprotein product of *T. saginata*. The antiHP10 had 72% sensitivity in CSF of confirmed NCC patients. It confirmed the presence of extra-parenchymal, but not parenchymal, NCC and worked better with CSF sample than serum sample. Another MoAb against cyst fluid of *T. solium* are IF11, IgG1 isotype had 82% sensitivity whereas same group of IgG MoAb was 4F8 used to detect the circulating Ag in subcutaneous cysticercosis patient serum (Harrison et al. 1989; Chang-Yuan, Hong-Hua, and Ling-Yun 1992).

3.1.5. Direct ELISA

In Direct ELISA antigens are directly coated on the solid surface of ELISA plate and are detected by enzyme conjugated Ab. This method is much faster and with low errors than other ELISA methods as in this method fewer steps are involved and this also lowers the cross reactivity. Rossa N et al. 1886 divided the CSF and serum samples in six groups of chronic meningitis, parenchymal cyst, intraventricular cyst, and chronic meningitis plus parenchymal cyst correspond to active NCC whereas parenchymal granulomas and subarachnoid fibrosis with hydrocephalus correspond to inactive from of NCC. They also took CSF from patients with other neurological disorders as negative control. The Ags were coated directly to the ELISA plate and detected by phosphatase conjugated anti-human IgM antibody. They got a sensitivity of 50% and 87% and 70% and 95% specificity, respectively with serum and CSF (Rosas, Sotelo, and Nieto 1986).

3.1.6. Antibody ELISA

The Ab-ELISA is used to detect the circulating Ag specific antibodies and reflects the exposure of parasite but may not confer the infection establishment or current infection. It uses mainly semi purified cyst fluid Ags (Garcia et al. 2018). The presence of immunoglobin class against Ag in serum marked for humoral response. The immunoglobulin IgG, IgM, IgE, IgA and IgD were detected in NCC patient's serum in decreasing order, while IgG, IgA, and IgE were detected in CSF samples. The highest diagnostic potential for ELISA was with IgG, having 93.2% specificity and 88.5% sensitivity while with IgM (96.6% and 65.4%), IgE (95.2% and 25.6%), and IgA (97.6% and 25.6%) (Odashima, Takayanagui, and De Castro Figueiredo 2002). Atluri et al. 2009 used excretory secretory and crude lysate Ag for the detection of circulating Ab detection in patient's serum and urine samples and found 30.4%, 63.2% and 38.4% sensitivity with serum sample respectively whereas with 47.2%, 44% and 46.6% respectively with urine samples.

3.1.7. DOT ELISA

In Dot ELISA, instead of ELISA plate a nitrocellulose (NC) membrane is used. Ags are spotted on NC at specific place and the other place of the membrane blocked with other proteins such BSA or milk to reduce false positive results. The NC membrane is then incubated with patient's serum or CSF sample. If the sample contains Abs specific to Ags, Abs will bind with the coated Ag which are then visualised by conjugated secondary Abs. Biswas R et al. 2004 compared the sensitivity and specificity of the Plate ELISA with Dot ELISA in NCC serum and CSF samples and observed better results with Plate ELISA. Dot ELISA had 56.25% sensitivity and 83% specificity whereas Plate ELISA had 73.75% sensitivity and 98% specificity (Biswas, Parija, and Narayan 2004).

The major limitations for NCC ELISA are they only give the idea of infection and presence of parasite but still it is important to be considered that this will vary depending upon stage and number of parasites. The calcified stage may typically give antibody positive but found to be Ag negative due to persistence response. It also needs to be considered that, ELISA does not give the exact idea of parasite stage as of the brain imaging. The serum based ELISA are not reliable and taking CSF sample is not easy and ethically correct for NCC management.

3.1.8. Molecular Assay and New Age Tools

Molecular diagnostic tools are choice of the day for any infectious disease. Among them most commonly used are PCR and Real Time QPCR. As per the diagnostic guideline radioimaging is the confirmative criteria for NCC. These radioimaging tools have many challenges in endemic region; the cost of instruments, technical expertise and their availability are limiting factors. Over the past decades many Taenia antigens or cross species antigens have been tested for better diagnosis and there is ongoing search to look for better biomarkers for NCC. On the other hand, the serological tools have not been tested in endemic regions and they fall low on specificity and have issues of cross reactivity, and moreover the accepted serological tool EITB, antigen preparation with LLGP is labour intensive. These all problems ask for development of reliable and easy to use molecular-diagnostic tool.

3.1.9. Polymerase Chain Reaction

PCR is simple, sensitive, and popular method for the diagnosis of number of infections, it requires the knowledge of DNA sequence to design primers and amplify the DNA fragment.

Considering the high specificity and sensitivity of the PCR based methods and with low cost and lesser time than other neuro diagnostic tools it can be used as reliable adjunct. For the diagnosis of NCC direct PCR, qPCR, real time TaqMan PCR etc. have been developed. The molecular target for PCR can be Ribosomal nucleotide, Mitochondrial or repetitive sequence of *T. solium* genome (Flores et al. 2018).

For Diagnosis of the NCC, the presence of *T. solium* nucleic acid was examined by using primers against a highly repetitive element of *T. solium* genome Tsol9. (Genbank Accession number: U45987). The sensitivity and specificity are important to determine its clinical relevance in such technique and to minimize false positive results. The sensitivity was determined by serially diluting the DNA up to several fold in a known concentration for example from 1µg to 1fg.

Table 3. DNA markers used to diagnose *T. solium* infections

Nucleic Acid type	Molecular Technique	Nucleic acid marker	Sample type	Sensitivity	References
rDNA	DNA hybridization	pTS10		100ng	1
gDNA	DNA hybridization	HDP1	Taeniid egg	3ng	2
gDNA	DNA hybridization	HDP2		25ng	2
rDNA	PCR sequencing	28s rRNA	Proglottids	500ng	3
gDNA	DNA hybridization	pTsol9	Taeniid egg	50-100ng	4
mtDNA	PCR sequencing	12s rDNA	Parasite tissue	200-800ng	5
rDNA	PCR RFLP	5.8s rRNA and ITSs rRNA	Proglottids	100ng	6
mtDNA	PCR sequencing	CYTb and CO1	Proglottids and egg		7
gDNA	Nested PCR	Tsol31	Feces	100fg	8
mtDNA	Multiplex PCR	Valine tRNA	Feces		9
rDNA	Real time PCR	ITS1/FAM		1fg	10
rDNA	Taq Man OCR	ITS1/ Cy5 probe			11
gDNA	Semi-nested PCR		CSF	174attogram	12
gDNA	TaqMan	pTsol9	CSF		13

Almedia et al. 2006 had obtained 67% detection rate when they used the pure CSF sample for PCR, whereas in concentrated CSF sample the detection level increased to 96.7% and sensitivity up to 10fg of DNA sample (Almeida et al. 2006).

Myata et al. 2008 used semi-nested PCR to amplify DNA samples extracted from stool sample of patient's having Taeniasis and found its 97% sensitivity and 100% specificity against the Tsol31 an oncosphere specific protein of *T. solium*. This assay does not give positive result with the *T. asiatica* and *T. saginata*. This method enhances the sensitivity and specificity of the detection of DNA sequence. Hernaddez et al. 2008 used semi-nested PCR (for its better specificity) for the detection of *T. solium* DNA in human CSF samples. They were able to detect the DNA sequence up to 174 atto-grams and had100% sensitivity (Hernández et al. 2008).

3.1.10. TaqMan PCR

The basic principle of TaqMan relies on the basic PCR method with a probe. The probe is a complementary sequence of the target gene and intends to hybridise with targeted nucleic acid sequence. It has a marginally higher annealing temperature than primers so that it will be hybridized with targeted nucleic acid when the polymerisation begins. These TaqMan probes are labelled with two different fluorescent dyes known as Reporter (R) and Quencher (Q) that fluoresce at different wavelengths. R is attached to 5'-end and Q is attached to 3'- end of the probe. When the probe is intact Q dye is in proximity of reporter dye and that results in no fluorescence but during polymerisation due to the 5' exonuclease activity of Taq DNA polymerase it cleaves off the probe and thus both R and Q dyes are released from the probe and Q dye is not in the close proximity of R and thus R fluorescences. Higher the amplicon higher the fluorescent level. Examples of R dye includes FAM or VIC, and for Q dye includes TAMARA I, which are used in combination for TaqMan Probe. H. Year et al. 2011 used this TaqMan PCR to detect *T. solium* DNA in CSF and serum samples. They used pTsol9 repetitive element as a target sequence. This method gave 100% specificity and 92% sensitivity with NCC patient's CSF samples. But this method was not able to detect the presence of DNA in blood and in active subarachnoid or ventricular cyst the sensitivity was 83-92% (Yera et al. 2011). They used the following primer for their assay:

Primer (TSF, 5'-CGATATTGAGCTAAGCTT-3'; TSR reverse complement, 5'-AGGAGGCCAGTTGCCTAGC-3').

3.1.11. Quantitative PCR (qPCR)

The real time quantitative PCR is used as a most reliable and attractive method for developing biomarkers for diagnosis purpose. This real time PCR includes a DNA binding dye such as SYBR green which binds between the dsDNA major groove but not with the ssDNA and emits fluorescence upon binding. The higher the amplification of DNA the higher the fluorescence level. The quantification of DNA is performed by making a standard curve using a range of known amounts of DNA concentration. O'Connell et al. 2020 had reported a new candidate of *T. solium* repeats TsolR13 are superior in terms of sensitivity and specificity than previously found repeats. The sensitivity at 240 atto-gram was 97.3% with analytical specificity of 100% and is a good candidate for NCC diagnosis by qPCR, In the clinical stage of active subarachnoid or ventricular NCC, the sensitivity with CSF and plasma was 100% and 81.35%, respectively and it gave specificity of 94.4% and 80.7% in CSF and Plasma of cured patients (Praet et al. 2013; O'Connell et al. 2020).

3.1.12. Multiplex PCR

Multiplex PCR uses a set of primers for different genes to amplify in a single run. It is important to determine whether these PCR methods discriminate between species or not. González et al. developed a multiplex PCR against the HDP2 gene of Taenia species. This one step PCR can differentiate the *T. solium, T. asiatica* and *T. saginata* infection and can also identify the coexistence of the infection. Yamasaki, et al. 2004 used the cytochrome C oxidase subunit 1 gene as a target for multiplex PCR. This assay can differentiate between the human Taeniidae or cysticercosis and can differentiate between the two genotypes of *taenia also* (Gonzlez et al. 2010).

3.1.13. Lymphocyte Transformation Test (LTT)

Peripheral blood mononuclear cells (PBMC) contain mature lymphocytes that circulate in the blood, they comprise of T-cells, NK cells and B cells. LTT has been used for the diagnosis of various autoimmune and allergic diseases. LTT provided best result for the diagnosis of NCC infection among patient with seizure and without seizure. For the detection of NCC infection, freshly isolated PBMCs were incubated with cyst fluid antigens in a 96 well plates and to detect the proliferation rate, radioactive tritiated Thymidine was used. The radio activity was measured using beta liquid scintillation. It was reported it had 93.75% sensitivity and 96.2% specificity while it had relatively low sensitivity (87.5%) for single cyst infection detection (Verma et al. 2010; A. Prasad et al. 2008). Though this technique is very sensitive and specific but it needs skilled manpower and needs to be tested in community settings.

3.1.14. Loop Mediated Isothermal Amplification (LAMP)

LAMP is a thermal amplification of DNA segment in a single constant temperature. As is the case with conventional PCR, it does not require a thermal cycler and can be performed in a single tube. In LAMP assay there are four primers involved for the amplification of a targeted nucleic acid using DNA polymerase. The four distinct primers (known as inner primers and outer primers) specifically designed to recognise six specific sequence on the target DNA. Each inner primer contains two different sequence, one will bind to the complementary region at the 3' terminal of sense sequence while another will bind with the 5' terminal inner region of the same chain. These two sequences have a spacer region such as TTTTT. Thus, using these primers the elongation reaction amplifies the targeted DNA by strand displacement activity using DNA polymerase. Characteristically, the four-specific nature of primers used in LAMP assay make it a highly specific assay. This assay can amplify a low copy of the DNA to 10^9 copies in less than an hour. Detection of the amplicon can be done by naked eye using magnesium pyrophosphate as a by-product which makes the solution turbid or using calcein, hydroxynapthol (Notomi et al. 2000). However, detection by the presence of turbidity or using nonspecific dye can give false positive result which can be avoided by using fluoresceine dye such as and SYBR green in a real time fluorometer. By using real time LAMP assay to amplify mitochondrial cox1 gene of *Taenia solium*, Goyal et al. 2019 were able to detect up to 1fg of cysticercal DNA of the patient's blood sample and found its sensitivity to 74% in NCC patients while 86.7% and 71.8% sensitivity in patients with extraparenchymal and intraparenchymal brain cyst, respectively. With single and multiple cyst the sensitivity was 73.5% and 74.5% respectively (Goyal et al. 2020).

The LAMP assay can be a promising tool for the diagnosis of NCC, as it can be done using a single tube reaction in an isothermal condition which does not require sophisticated instruments, it can amplify very few copies of DNA which makes it a highly sensitive method for diagnosis purpose. The limited use of instrument and laboratory make this assay cost friendly and can be a used as an alternative method for the diagnosis of NCC in poor or developing countries or in the ruler areas where the clinical laboratories don't have much facilities.

3.1.15. Magnetic Resonance Spectroscopy (MRS)

MRS is a non-invasive technique that allows clinicians to determine the quantity of different metabolites in diseased and healthy conditions. In case of CNS infections or other neurodegenerative conditions, brain metabolites concentration changes and MRS technique is used to measure quantity of selected metabolites such as N-acetyl aspartate (NAA), creatine, choline, lactate, gamma-aminobutyric acid (GABA), glutamate and myo-inositol(Buonocore and Maddock 2015; Tognarelli et al. 2015). In case of parenchymal NCC, MRS is used to determine the concentration of different metabolites. It was reported that in NCC infection there was upregulation of certain metabolites such as lactate, alanine, succinate, pyruvate and choline. The MR spectra further supported the MRI scan images and confirmed the presence of cystic lesions (Pandit et al. 2001; Chawla et al. 2004).

3.1.16. miRNA

MicroRNA (miRNA) is a single stranded, non-coding, endogenous short RNA which plays important role in post transcription regulation. Helminths secretes miRNA during infection which is detected in human blood. It makes them a promising biomarker for diagnosis of helminth infections. The relationship of miRNA and *T. solium* is little known and it has never been used as a biomarker for NCC. However, as a novel regulator in infection and disease there is need for detailed research about miRNA in NCC and that may give lead for potential markers for the diagnosis of NCC (Gutierrez-Loli et al. 2017).

3.1.17. Exosome's Detection

Exosomes are extracellular vesicles generated by the cells of host and pathogens too, and they carry nuclei acids, proteins, lipids and metabolites. The main function of exosomes is in cell-to-cell communication. During the course of infection, helminths secrete their exosomes to regulate the immune response and found in host body fluids. The concentration of secreted exosomes in the body fluids is very less, and imposed major limitation to understand its role in infection or use for diagnostic purpose (Lin et al. 2015; Kanninen et al. 2016).

CONCLUSION

Accurate diagnosis of NCC is crucial for better management of patient, but the existing guideline is complex to come up with a definitive diagnosis of NCC specially in the absence of readily available low cost clinically acceptable diagnostic tool. The accepted diagnostic tools i.e., MRI and CT are not available in endemic areas readily, specially to rural population, hence there is an effort going on to come up with reliable serological or molecular tools. Several promising tools had been tested in recent past but their acceptance by clinical community is low, as most of these have been evaluated in limited population. However, with the determination of World Health Organisation in eradicating the disease by 2030, better effort to develop reliable and cost effective diagnostic tool is going on. Development of a point of care device or tool based on lateral flow technologies will be useful for epidemiological studies or to assist health care provider at primary care centres in filed settings.

REFERENCES

Almeida, C. R., E. P. Ojopi, C. M. Nunes, L. R. Machado, O. M. Takayanagui, J. A. Livramento, R. Abraham, W. F. Gattaz, A. J. Vaz, and E. Dias-Neto. 2006. "Taenia Solium DNA Is Present in the Cerebrospinal Fluid of Neurocysticercosis Patients and Can Be Used for Diagnosis." *European Archives of Psychiatry and Clinical Neuroscience* 256 (5): 307–10. https://doi.org/10.1007/s00406-006-0612-3.

Arora, Naina, Rimanpreet Kaur, Farhan Anjum, Shweta Tripathi, Amit Mishra, Rajiv Kumar, and Amit Prasad. 2019. "Neglected Agent Eminent Disease: Linking Human Helminthic Infection, Inflammation, and Malignancy." *Frontiers in Cellular and Infection Microbiology*. Frontiers Media S. A. https://doi.org/10.3389/fcimb.2019.00402.

Arora, Naina, Rimanpreet Kaur, Suraj Singh Rawat, Ankur Kumar, Aloukick Kumar Singh, Shweta Tripathi, Amit Mishra, Gagandeep Singh, and Amit Prasad. 2020. "Evaluation of Taenia Solium Cyst Fluid-Based Enzyme Linked Immunoelectro Transfer Blot for Neurocysticercosis Diagnosis in Urban and Highly Endemic Rural Population of North India." *Clinica Chimica Acta* 508: 16–21. https://doi.org/10.1016/j.cca.2020.05.006.

Arora, Naina, Shweta Tripathi, Reshma Sao, Prosenjit Mondal, Amit Mishra, and Amit Prasad. 2018. "Molecular Neuro-Pathomechanism of Neurocysticercosis: How Host Genetic Factors Influence Disease Susceptibility." *Molecular Neurobiology*. Humana Press Inc. https://doi.org/10.1007/s12035-016-0373-6.

Atluri, Subba Rao V., P. Singhi, N. Khandelwal, and N. Malla. 2009. "Neurocysticercosis Immunodiagnosis Using Taenia Solium Cysticerci Crude Soluble Extract, Excretory Secretory and Lower Molecular Mass Antigens in Serum and Urine Samples of Indian Children." *Acta Tropica* 110 (1): 22–27. https://doi.org/10.1016/j.actatropica.2008.12.004.

Biswas, Rakhi, S. C. Parija, and S. K. Narayan. 2004. "Dot-ELISA for the Diagnosis of Neurocysticercosis." *Revista Do Instituto de Medicina Tropical de Sao Paulo* 46 (5): 249–52. https://doi.org/10.1590/s0036-46652004000500003.

Braga, Flavio T., Antonio J. Da Rocha, Ricardo B. Fonseca, Chantal Frigon, Dennis W. W. Shaw, Susan Heckbert, and Ed Weinberger. 2005. "Supplemental Oxygen Concentration and Increased Signal Intensity of Cerebrospinal Fluid on FLAIR MR Images [2] (Multiple Letters)." *Radiology*. Radiological Society of North America . https://doi.org/10.1148/radiol.2353041898.

Brutto, O. H. Del, T. E. Nash, A. C. White, V. Rajshekhar, P. P. Wilkins, G. Singh, C. M. Vasquez, P. Salgado, R. H. Gilman, and H. H. Garcia. 2017. "Revised Diagnostic Criteria for Neurocysticercosis." *Journal of the Neurological Sciences*. Elsevier B. V. https://doi.org/10.1016/j.jns. 2016.11.045.

Brutto, Oscar H. Del, Gianfranco Arroyo, Victor J. Del Brutto, Mauricio Zambrano, and Héctor H. García. 2017. "On the Relationship between Calcified Neurocysticercosis and Epilepsy in an Endemic Village: A Large-Scale, Computed Tomography–Based Population Study in Rural Ecuador." *Epilepsia* 58 (11): 1955–61. https://doi.org/10.1111/epi. 13892.

Brutto, Oscar H. Del, V. Rajshekhar, A. C. White, V. C. W. Tsang, T. E. Nash, O. M. Takayanagui, P. M. Schantz, et al. 2001. "Proposed Diagnostic Criteria for Neurocysticercosis." *Neurology*. Lippincott Williams and Wilkins. https://doi.org/10.1212/WNL.57.2.177.

Buonocore, Michael H., and Richard J. Maddock. 2015. "Magnetic Resonance Spectroscopy of the Brain: A Review of Physical Principles and Technical Methods." *Reviews in the Neurosciences*. Walter de Gruyter GmbH. https://doi.org/10.1515/revneuro-2015-0010.

Cangalaya, Carla, Javier A. Bustos, Juan Calcina, Ana Vargas-Calla, Diego Suarez, Armando E. Gonzalez, Juan Chacaltana, et al. 2016. "Perilesional Inflammation in Neurocysticercosis - Relationship Between Contrast-Enhanced Magnetic Resonance Imaging, Evans Blue Staining and Histopathology in the Pig Model." *PLoS Neglected Tropical Diseases* 10 (7). https://doi.org/10.1371/journal.pntd. 0004869.

Chang-Yuan, W., Z. Hong-Hua, and G. Ling-Yun. 1992. "A MAb-Based ELISA for Detecting Circulating Antigen in CSF of Patients with Neurocysticercosis." *Hybridoma* 11 (6): 825–27. https://doi.org/10. 1089/hyb.1992.11.825.

Chawla, Sanjeev, Rakesh K. Gupta, Nuzhat Husain, Monika Garg, Rajesh Kumar, and Sunil Kumar. 2004. "Prediction of Viability of Porcine Neurocysticercosis with Proton Magnetic Resonance Spectroscopy:

Correlation with Histopathology." *Life Sciences* 74 (9): 1081–92. https://doi.org/10.1016/j.lfs.2003.07.031.

Chung, Joon Yong, Young Yil Bahk, Sun Huh, Shin Yong Kong, Yoon Kong, and Seung Yull Cho. 1999. "A Recombinant 10-KDa Protein of Taenia Solium Metacestodes Specific to Active Neurocysticercosis." *Journal of Infectious Diseases* 180 (4): 1307–15. https://doi.org/10.1086/315020.

Debacq, Gabrielle, Luz M. Moyano, Héctor H. Garcia, Farid Boumediene, Benoit Marin, Edgard B. Ngoungou, and Pierre Marie Preux. 2017. "Systematic Review and Meta-Analysis Estimating Association of Cysticercosis and Neurocysticercosis with Epilepsy." *PLoS Neglected Tropical Diseases* 11 (3). https://doi.org/10.1371/journal.pntd.0005153.

Espinoza B, Ruiz-Palacios G, Tovar A et al. Characterization by enzyme-linked immunosorbent assay of the humoral immune response in patients with neurocysticercosis and its application in immunodiagnosis. *J Clin Microbiol*. 1986; 24: 536–541.

Feldman, Miriam, Augustín Plancarte, Miguel Sandoval, Marianna Wilson, and Ana Flisser. 1990. "Comparison of Two Assays (EIA and EITB) and Two Samples (Saliva and Serum) for the Diagnosis of Neurocysticercosis." *Transactions of the Royal Society of Tropical Medicine and Hygiene* 84 (4): 559–62. https://doi.org/10.1016/0035-9203(90)90040-L.

Flisser A, Tarrab R, Willms K, Larralde C. Immunoelectrophoresis and double immunodiffusion in the diagnosis of human cerebral cysticercosis; *Arch. Invest. Med*. (Mex). 1975; 6:1–12.

Flores, María D., Luis M. Gonzalez, Carolina Hurtado, Yamileth Monje Motta, Cristina Domínguez-Hidalgo, Francisco Jesús Merino, María J. Perteguer, and Teresa Gárate. 2018. "HDP2: A Ribosomal DNA (NTS-ETS) Sequence as a Target for Species-Specific Molecular Diagnosis of Intestinal Taeniasis in Humans." *Parasites and Vectors* 11 (1): 117. https://doi.org/10.1186/s13071-018-2646-6.

Frigon, Chantal, Dennis W. W. Shaw, Susan R. Heckbert, Edward Weinberger, and David S. Jardine. 2004. "Supplemental Oxygen

Causes Increased Signal Intensity in Subarachnoid Cerebrospinal Fluid on Brain FLAIR MR Images Obtained in Children during General Anesthesia." *Radiology* 233 (1): 51–55. https://doi.org/10.1148/radiol.2331031375.

Garcia HH, Del Brutto OH. Imaging findings in neurocysticercosis. *Acta Tropica*. 2003; 87: 71-78.

Garcia, Hector H., Seth E. O'Neal, John Noh, Sukwan Handali, Robert H. Gilman, Armando E. Gonzalez, Victor C. W. Tsang, et al. 2018. "Laboratory Diagnosis of Neurocysticercosis (Taenia Solium)." *Journal of Clinical Microbiology* 56 (9). https://doi.org/10.1128/JCM.00424-18.

Garcia, Hector H., Theodore E. Nash, and Oscar H. Del Brutto. 2014. "Clinical Symptoms, Diagnosis, and Treatment of Neurocysticercosis." *The Lancet Neurology*. Lancet Publishing Group. https://doi.org/10.1016/S1474-4422(14)70094-8.

Gasparetto, Emerson Leandro, Soniza Alves-Leon, Flavio Sampaio Domingues, João Thiago Frossard, Selva Paraguassu Lopes, and Jorge Marcondes de Souza. 2016. "Neurocysticercosis, Familial Cerebral Cavernomas and Intracranial Calcifications: Differential Diagnosis for Adequate Management." *Arquivos de Neuro-Psiquiatria*. Associacao Arquivos de Neuro-Psiquiatria. https://doi.org/10.1590/0004-282X20160054.

Gonzalez D, Rodriguez-Carbajal J, Aluja A, Flisser A. Cerebral cysticercosis in pigs studied by computed tomography and necroscopy. *Vet Parasitol*. 1987; 26: 55-69.

Gonzlez, Luis M., Begõa Bailo, Elizabeth Ferrer, Maria D. Fernandez García, Leslie Js Harrison, Michael Re Parkhouse, Donald P. McManus, and Teresa Grate. 2010. "Characterization of the Taenia Spp HDP2 Sequence and Development of a Novel PCR-Based Assay for Discrimination of Taenia Saginata from Taenia Asiatica." *Parasites and Vectors* 3 (1): 51. https://doi.org/10.1186/1756-3305-3-51.

Goyal, Gunjan, Anil Chandra Phukan, Masaraf Hussain, Vivek Lal, Manish Modi, Manoj Kumar Goyal, and Rakesh Sehgal. 2020. "Sorting out Difficulties in Immunological Diagnosis of

Neurocysticercosis: Development and Assessment of Real Time Loop Mediated Isothermal Amplification of Cysticercal DNA in Blood." *Journal of the Neurological Sciences* 408 (January): 116544. https://doi.org/10.1016/j.jns.2019.116544.

Greene, R. M., K. Hancock, P. P. Wilkins, and V. C. W. Tsang. 2000. "Taenia Solium: Molecular Cloning and Serologic Evaluation of 14- and 18-KDa Related, Diagnostic Antigens." *Journal of Parasitology* 86 (5): 1001–7. https://doi.org/10.1645/0022-3395(2000)086[1001: tsmcas]2.0.co;2.

Grogl, Max, John J. Estrada, Gene MacDonald, and Raymond E. Kuhn. 1985. "Antigen-Antibody Analyses in Neurocysticercosis." *The Journal of Parasitology* 71 (4): 433. https://doi.org/10.2307/3281534.

Gutierrez-Loli, Renzo, Miguel A. Orrego, Oscar G. Sevillano-Quispe, Luis Herrera-Arrasco, and Cristina Guerra-Giraldez. 2017. "MicroRNAs in Taenia Solium Neurocysticercosis: Insights as Promising Agents in Host-Parasite Interaction and Their Potential as Biomarkers." *Frontiers in Microbiology* 8 (SEP): 1905. https://doi.org/10.3389/fmicb.2017.01905.

Harrison, L. J. S., G. W. P. Joshua, S. H. Wright, and R. M. E. Parkhouse. 1989. "Specific Detection of Circulating Surface/Secreted Glycoproteins of Viable Cysticerci in *Taenia Saginata* Cysticercosis." *Parasite Immunology* 11 (4): 351–70. https://doi.org/10.1111/j.1365-3024.1989.tb00673.x.

Hernández, M., L. M. Gonzalez, A. Fleury, B. Saenz, R. M. E. Parkhouse, L. J. S. Harrison, T. Garate, and E. Sciutto. 2008. "Neurocysticercosis: Detection of Taenia Solium DNA in Human Cerebrospinal Fluid Using a Semi-Nested PCR Based on HDP2." *Annals of Tropical Medicine and Parasitology* 102 (4): 317–23. https://doi.org/10.1179/136485908 X278856.

Hernández-González, Ana, John Noh, María Jesús Perteguer, Teresa Gárate, and Sukwan Handali. 2017. "Comparison of T24H-His, GST-T24H and GST-Ts8B2 Recombinant Antigens in Western Blot, ELISA and Multiplex Bead-Based Assay for Diagnosis of

Neurocysticercosis." *Parasites and Vectors* 10 (1). https://doi.org/10.1186/s13071-017-2160-2.

Ito A, Plancarte A, Ma L. Novel antigens for neurocysticercosis: simple method for preparation and evaluation for serodiagnosis. *Am. J. Trop. Med. Hyg.* 1998; 59: 291–294.

Ito, Akira, Hiroshi Yamasaki, Minoru Nakao, Yasuhito Sako, Munehiro Okamoto, Marcello O. Sato, Kazuhiro Nakaya, et al. 2003. "Multiple Genotypes of Taenia Solium - Ramifications for Diagnosis, Treatment and Control." In *Acta Tropica*, 87:95–101. Elsevier. https://doi.org/10.1016/S0001-706X(03)00024-X.

Kanninen, Katja M., Nea Bister, Jari Koistinaho, and Tarja Malm. 2016. "Exosomes as New Diagnostic Tools in CNS Diseases." *Biochimica et Biophysica Acta - Molecular Basis of Disease* 1862 (3): 403–10. https://doi.org/10.1016/j.bbadis.2015.09.020.

Kimura-Hayama, Eric T., Jesús A. Higuera, Roberto Corona-Cedillo, Laura Chávez-Macías, Anamari Perochena, Laura Yadira Quiroz-Rojas, Jesús Rodríguez-Carbajal, and José L. Criales. 2010a. "Neurocysticercosis: Radiologic-Pathologic Correlation." *Radiographics* 30 (6): 1705–19. https://doi.org/10.1148/rg.306105522.

———. 2010b. "Neurocysticercosis: Radiologic-Pathologic Correlation." *Radiographics* 30 (6): 1705–19. https://doi.org/10.1148/rg.306105522.

———. 2010c. "Neurocysticercosis: Radiologic-Pathologic Correlation." *Radiographics* 30 (6): 1705–19. https://doi.org/10.1148/rg.306105522.

Kuehnast, M., S. Andronikou, L. T. Hlabangana, and C. N. Menezes. 2020. "Imaging of Neurocysticercosis and the Influence of the Human Immunodeficiency Virus." *Clinical Radiology* 75 (1): 77.e1-77.e13. https://doi.org/10.1016/j.crad.2019.08.001.

Lin, Jin, Jing Li, Bo Huang, Jing Liu, Xin Chen, Xi Min Chen, Yan Mei Xu, Lin Feng Huang, and Xiao Zhong Wang. 2015. "Exosomes: Novel Biomarkers for Clinical Diagnosis." *Scientific World Journal* 2015 (January). https://doi.org/10.1155/2015/657086.

Lucato, Leandro Tavares, M. S. Guedes, J. R. Sato, L. A. Bacheschi, L. R. Machado, and C. C. Leite. 2007. "The Role of Conventional MR Imaging Sequences in the Evaluation of Neurocysticercosis: Impact on

Characterization of the Scolex and Lesion Burden." *American Journal of Neuroradiology* 28 (8): 1501–4. https://doi.org/10.3174/ajnr.A0623.

Madrazo, I., J. A. Garcia Renteria, M. Sandoval, and F. J. Lopez Vega. 1983. "Intraventricular Cysticercosis." *Neurosurgery* 12 (2): 148–52. https://doi.org/10.1227/00006123-198302000-00003.

Mehemed, Taha M., Yasutaka Fushimi, Tomohisa Okada, Akira Yamamoto, Mitsunori Kanagaki, Aki Kido, Koji Fujimoto, Naotaka Sakashita, and Kaori Togashi. 2014. "Dynamic Oxygen-Enhanced MRI of Cerebrospinal Fluid." Edited by Friedemann Paul. *PLoS ONE* 9 (6): e100723. https://doi.org/10.1371/journal.pone.0100723.

Montano, S. M., M. V. Villaran, L. Ylquimiche, J. J. Figueroa, S. Rodriguez, C. T. Bautista, A. E. Gonzalez, V. C. W. Tsang, R. H. Gilman, and H. H. Garcia. 2005. "Neurocysticercosis: Association between Seizures, Serology, and Brain CT in Rural Peru." *Neurology* 65 (2): 229–34. https://doi.org/10.1212/01.wnl.0000168828.83461.09.

Ndimubanzi, Patrick C., Hélène Carabin, Christine M. Budke, Hai Nguyen, Ying-Jun Qian, Elizabeth Rainwater, Mary Dickey, Stephanie Reynolds, and Julie A. Stoner. 2010. "A Systematic Review of the Frequency of Neurocyticercosis with a Focus on People with Epilepsy." Edited by Pierre-Marie Preux. *PLoS Neglected Tropical Diseases* 4 (11): e870. https://doi.org/10.1371/journal.pntd.0000870.

Notomi, T., H. Okayama, H. Masubuchi, T. Yonekawa, K. Watanabe, N. Amino, and T. Hase. 2000. "Loop-Mediated Isothermal Amplification of DNA." *Nucleic Acids Research* 28 (12): e63. https://doi.org/10.1093/nar/28.12.e63.

O'Connell, Elise M, Sarah Harrison, Eric Dahlstrom, Theodore Nash, and Thomas B Nutman. 2020. "A Novel, Highly Sensitive Quantitative Polymerase Chain Reaction Assay for the Diagnosis of Subarachnoid and Ventricular Neurocysticercosis and for Assessing Responses to Treatment." *Clinical Infectious Diseases* 70 (9): 1875–81. https://doi.org/10.1093/cid/ciz541.

Obregón-Henao, Andrés, Diana P. Londoño, Diana I. Gómez, Judith Trujillo, Judy M. Teale, and Blanca I. Restrepo. 2003. "*In Situ* Detection of Antigenic Glycoproteins in Taenia Solium

Metacestodes." *Journal of Parasitology* 89 (4): 726–32. https://doi.org/10.1645/GE-3046.

Odashima, Newton Satoru, Osvaldo Massaiti Takayanagui, and José Fernando De Castro Figueiredo. 2002. "Enzyme Linked Immunosorbent Assay (ELISA) for the Detection of IgG, IgM, IgE and IgA against Cysticercus Cellulosae in Cerebrospinal Fluid of Patients with Neurocysticercosis." *Arquivos de Neuro-Psiquiatria* 60 (2 B): 400–405. https://doi.org/10.1590/S0004-282X2002000300012.

Pandit, Sameer, Alexander Lin, Helmuth Gahbauer, Claudia R. Libertin, and Barbaros Erdogan. 2001. "MR Spectroscopy in Neurocysticercosis." *Journal of Computer Assisted Tomography* 25 (6): 950–52. https://doi.org/10.1097/00004728-200111000-00019.

Pooley, Robert A. 2005. "Fundamental Physics of MR Imaging." *Radiographics* 25 (4): 1087–99. https://doi.org/10.1148/rg.254055027.

Prabhakaran V, Rajshekhar V, Murrell KD, Oommen A. Taenia solium metacestode glycoproteins as diagnostic antigens for solitary cysticercus granuloma in Indian patients. *Trans R Soc Trop Med Hyg.* 2004; 98:478-484.

Praet, Nicolas, Jaco J. Verweij, Kabemba E. Mwape, Isaac K. Phiri, John B. Muma, Gideon Zulu, Lisette van Lieshout, et al. 2013. "Bayesian Modelling to Estimate the Test Characteristics of Coprology, Coproantigen ELISA and a Novel Real-Time PCR for the Diagnosis of Taeniasis." *Tropical Medicine & International Health* 18 (5): 608–14. https://doi.org/10.1111/tmi.12089.

Prasad, Amit, Kashi Nath Prasad, Abhisek Yadav, Rakesh Kumar Gupta, Sunil Pradhan, Sanjeev Jha, Mukesh Tripathi, and Mazhar Husain. 2008. "Lymphocyte Transformation Test: A New Method for Diagnosis of Neurocysticercosis." *Diagnostic Microbiology and Infectious Disease* 61 (2): 198–202. https://doi.org/10.1016/j.diagmicrobio.2007.12.016.

Prasad, Kashi N., Amit Prasad, Rakesh K. Gupta, Kavindra Nath, Sunil Pradhan, Mukesh Tripathi, and Chandra M. Pandey. 2009. "Neurocysticercosis in Patients with Active Epilepsy from the Pig Farming Community of Lucknow District, North India." *Transactions* of the Royal Society of Tropical Medicine and Hygiene 103 (2): 144–50. https://doi.org/10.1016/j.trstmh.2008.07.015.

Prasad, Kashi Nath, Amit Prasad, Avantika Verma, and Aloukick Kumar Singh. 2008. "Human Cysticercosis and Indian Scenario: A Review." *Journal of Biosciences*. Springer. https://doi.org/10.1007/s12038-008-0075-y.

Rajshekhar, V., M. Venkat Raghava, V. Prabhakaran, A. Oommen, and J. Muliyil. 2006. "Active Epilepsy as an Index of Burden of Neurocysticercosis in Vellore District, India." *Neurology* 67 (12): 2135–39. https://doi.org/10.1212/01.wnl.0000249113.11824.64.

Romo, Matthew L., Arturo Carpio, R. Michael E. Parkhouse, María Milagros Cortéz, and Richar Rodríguez-Hidalgo. 2018. "Comparison of Complementary Diagnostic Tests in Cerebrospinal Fluid and Serum for Neurocysticercosis." *Heliyon* 4 (12): e00991. https://doi.org/10.1016/j.heliyon.2018.e00991.

Rosas, Norma, Julio Sotelo, and Dionisio Nieto. 1986. "ELISA in the Diagnosis of Neurocysticercosis." *Archives of Neurology* 43 (4): 353–56. https://doi.org/10.1001/archneur.1986.00520040039016.

Singhi, Pratibha. 2011. "Neurocysticercosis." *Therapeutic Advances in Neurological Disorders* 4 (2): 67–81. https://doi.org/10.1177/1756285610395654.

Suh, D. C., K. H. Chang, M. H. Han, S. R. Lee, M. C. Han, and C. W. Kim. 1989. "Unusual MR Manifestations of Neurocysticercosis." *Neuroradiology* 31 (5): 396–402. https://doi.org/10.1007/BF00343863.

Tognarelli, Joshua M., Mahvish Dawood, Mohamed I. F. Shariff, Vijay P. B. Grover, Mary M. E. Crossey, I. Jane Cox, Simon D. Taylor-Robinson, and Mark J. W. McPhail. 2015. "Magnetic Resonance Spectroscopy: Principles and Techniques: Lessons for Clinicians." *Journal of Clinical and Experimental Hepatology*. Elsevier B. V. https://doi.org/10.1016/j.jceh.2015.10.006.

Torgerson, Paul R., Brecht Devleesschauwer, Nicolas Praet, Niko Speybroeck, Arve Lee Willingham, Fumiko Kasuga, Mohammad B. Rokni, et al. 2015. "World Health Organization Estimates of the Global and Regional Disease Burden of 11 Foodborne Parasitic

Diseases, 2010: A Data Synthesis." Edited by Lorenz von Seidlein. *PLOS Medicine* 12 (12): e1001920. https://doi.org/10.1371/journal.pmed.1001920.

Tsang, V. C. W., J. A. Brand, and A. E. Boyer. 1989. "An Enzyme-Linked Immunoelectrotransfer Blot Assay and Glycoprotein Antigens for Diagnosing Human Cysticercosis (*Taenia Solium*)." *Journal of Infectious Diseases* 159 (1): 50–59. https://doi.org/10.1093/infdis/159.1.50.

Venkat, Bargavee, Neeti Aggarwal, Sushma Makhaik, and Ramgopal Sood. 2016. "A Comprehensive Review of Imaging Findings in Human Cysticercosis." *Japanese Journal of Radiology*. Springer Tokyo. https://doi.org/10.1007/s11604-016-0528-4.

Verma, Avantika, Kashi N. Prasad, Aloukick K. Singh, Kishan K. Nyati, Rakesh K. Gupta, and Vimal K. Paliwal. 2010. "Evaluation of the MTT Lymphocyte Proliferation Assay for the Diagnosis of Neurocysticercosis." *Journal of Microbiological Methods* 81 (2): 175–78. https://doi.org/10.1016/j.mimet.2010.03.001.

Watts, Nathaniel S., Monica Pajuelo, Taryn Clark, Maria-Cristina I. Loader, Manuela R. Verastegui, Charles Sterling, Jon S. Friedland, Hector H. Garcia, and Robert H. Gilman. 2014. "Taenia Solium Infection in Peru: A Collaboration between Peace Corps Volunteers and Researchers in a Community Based Study." Edited by Oliver Schildgen. *PLoS ONE* 9 (12): e113239. https://doi.org/10.1371/journal.pone.0113239.

Yera, H., D. Dupont, S. Houze, M. Ben M'Rad, F. Pilleux, A. Sulahian, C. Gatey, F. Gay Andrieu, and J. Dupouy-Camet. 2011. "Confirmation and Follow-up of Neurocysticercosis by Real-Time PCR in Cerebrospinal Fluid Samples of Patients Living in France." *Journal of Clinical Microbiology* 49 (12): 4338–40. https://doi.org/10.1128/JCM.05839-11.

Zhao, Jing-Long, Alexander Lerner, Zheng Shu, Xing-Jun Gao, and Chi-Shing Zee. 2015. "Imaging Spectrum of Neurocysticercosis." *Radiology of Infectious Diseases* 1 (2): 94–102. https://doi.org/10.1016/j.jrid.2014.12.001.

In: Neurocysticercosis　　　　　ISBN: 978-1-53619-791-4
Editor: Mark A. Chavez　　　　© 2021 Nova Science Publishers, Inc.

Chapter 3

TRENDS IN THE DIAGNOSIS OF HUMAN NEUROCYSTICERCOSIS: ISSUES AND CHALLENGES

Abhishek Mewara, MD and Nancy Malla, MD*
Department of Medical Parasitology,
Postgraduate Institute of Medical Education and Research,
Chandigarh, India

ABSTRACT

Human neurocysticercosis (NCC) is a disease caused when the larval stage (cysticerci) of *Taenia solium* lodge in the central nervous system. It is regarded as a significant public health problem in Asia, Africa and the Latin America. The infection is also increasingly being seen in more developed countries due to frequent travel and immigration from endemic areas. Taeniasis/cysticercosis is one of the neglected tropical diseases which are targeted for control by the WHO. The clinical diagnosis of NCC is presumptive and is usually substantiated by laboratory diagnostic

---

* Corresponding Author's E-mail: drmallanancy@gmail.com.

procedures. Radioimaging provides information regarding the number, size, and location of the cysts, but often is non-specific, mimicking other pathologies. Antibody, antigen, nucleic acid detection in body fluids and/or genotyping of the cysts usually substantiates and/or confirms the clinical diagnosis. However, all these techniques have their own merits and demerits, with varying sensitivity and specificity which depend upon the sample, assay and type of antigen used, besides clinical presentation including stage and location of the lesion. Therefore, application of a single conventional technique may not provide confirmatory diagnosis in all the clinically suspected patients. Moreover, molecular techniques are not usually available in endemic areas with limited resources and facilities. The combination of two or more techniques, when applied, may yield desirable sensitivity and specificity for the diagnosis and planning of treatment strategies. The detection of antibody response to crude soluble extract *T. solium* antigen by ELISA to achieve desirable sensitivity, followed by more specific technique with the use of lower molecular mass specific antigens to check specificity in seropositive individuals may serve useful purpose for diagnosis in endemic areas which have limited resources for molecular techniques. The diagnostic criteria for human NCC need to be defined specific to the endemic area, depending upon the clinical presentation, asymptomatic infections, clinically similar CNS pathologies and other associated factors in the specific regions. The antigen detection technique(s) in body fluids have been found helpful to assess the response to treatment on follow-up of patients. The advanced molecular techniques, nucleic acid detection by PCR and its modifications, isothermal nucleic acid amplification techniques such as LAMP, cell mediated methods, and advanced proteomics techniques are still under investigation and are yet to find a practical use in the diagnosis of NCC.

Keywords: cysticercosis, diagnosis, *Taenia solium*, neurocysticercosis

INTRODUCTION

Human neurocysticercosis (NCC) caused by ingestion of the eggs of *Taenia solium*, a cestode parasite, is a major public health problem in many endemic countries of Asia, Africa and Latin America [1, 2, 3]. The infection is also increasingly being observed in developed countries due to immigration from endemic areas as well as increased international travel [4]. Taeniasis/cysticercosis has been listed as a neglected tropical disease

(NTD) for control by the WHO [5]. The data on prevalence of the infection and/or disease cannot be accurately extrapolated due to biological, statistical and technical limitations, besides there being a large prevalence of asymptomatic cases, and insignificant immune response against the parasite in those infected; moreover, notification of the occurrence of this infection/disease is not mandatory. The prevalence of NCC is directly based on the prevalence of adult worm infestation in a population. A review revealed that 0.5% to 15% subjects in different geographical areas harbour the adult worms [6]. The WHO considers it a significant public health problem if *T. solium* infestation rate is >1% in a particular area. The latest reviews indicate that in endemic areas a significant number of patients suffering from epilepsy are due to NCC [7, 8, 9]. The increasing seropositivity to *T. solium* cysticercal antigen has been reported, based on the retrospective analysis of antibody positivity in clinically suspected NCC patients attending a tertiary care hospital in north India from 1994 to 2000, with higher seroprevalence in children [10].

Human beings are the only definitive host of *T. solium*, harbouring the adult worms in the small intestine. The parasitic ova are passed in the faeces of the definitive host and subsequently consumed by the intermediate host, the pig. The ova penetrate the intestinal wall of the intermediate host pig, gaining entrance into lymphatics and veins, further disseminating and maturing into larval forms in muscles and other organs of the body. However, humans can also act as an intermediate host by ingestion of eggs with contaminated vegetables or water, besides by autoinfection or by reverse peristalsis in the adult worm infected host. The cysts lodging in the central nervous system (CNS) lead to NCC, reported as the second most important cause of intracranial space occupying lesion after tuberculosis [1]. The clinical diagnosis is presumptive, as the symptoms and signs mimic a plethora of other clinical syndromes, thus necessitating the application of laboratory based diagnostic techniques. The identification of the cysts in biopsy samples is the confirmatory method to establish the diagnosis, however, it involves an invasive procedure, which may not be the preferred choice due to obvious limitations [2]. It is proposed that brain biopsy may be necessary in a patient with seizures and raised intracranial pressure with multiple small enhancing lesions and associated oedema in absence of calcifications, which mimic tuberculomas on neuroimaging [8].

Del Brutto et al. proposed the criteria for the diagnosis of NCC, based on absolute, major, minor and epidemiological criteria, which could indicate probable or definitive diagnosis [11]. It is suggested that these criteria need to be evaluated in India [12]. Further, on reviewing the proposed diagnostic criteria, it is documented that these criteria are being considered for diagnosis both in field and hospital settings, and in both endemic and non-endemic areas. The criteria were proposed to be considered worldwide, with scope of modifications if desired depending upon the most common pattern of disease presentation in a given area [13]. The revised guidelines [14], and clinical practice guidelines have been proposed to help clinicians and health workers for the management and diagnosis of human NCC [15]. However, a proper interpretation of these criteria permits two degrees of diagnostic certainty, definitive or probable. Moreover, the criteria may not serve useful purpose in all the endemic areas worldwide and modifications may be desired depending upon the clinical presentation and other associated factors in the specific regions.

The radioimaging techniques, detection of specific antibody and antigen in body fluids have been widely reported for the diagnosis of NCC with merits and demerits of each technique. The advanced techniques such as detection of specific nucleic acid sequences of the parasite genome, application of cell mediated methods and proteomic techniques are still under investigation. The radio-imaging techniques provide information regarding parasite cyst numbers, size, stage and location, however, many a times the findings are non-specific, mimicking other pathologies. The CT scan is considered the best tool for the diagnosis of intraparenchymal calcification and MRI for ventricular cysts [16], yet the inability of the radioimaging techniques to detect cysts located in the subarachnoid basal cisterns may pose a problem in the diagnosis of such conditions [7].

ANTIBODY DETECTION

A battery of immunodiagnostic techniques has been applied to detect antibody response to *T. solium* cysticerci antigens in serum and CSF samples to substantiate the clinical diagnosis and imaging reports. Enzyme-linked immunosorbent assay (ELISA) is widely reported for antibody detection with variable sensitivity and specificity from different geographical regions. Crude soluble *T. solium* cysticerci extract (CSE) antigen, purified antigenic fractions (PAF), excretory-secretory (ES) antigens and recombinant antigens have been used for the serological diagnosis of NCC with limitations of each antigen type.

Crude Soluble Extract Antigens

The reports, in general conclude that ELISA with the use of CSE antigen for antibody detection in serum yields high sensitivity but low specificity. On comparing the ELISA and indirect haemagglutination (IHA) technique, the antibody response to *T. solium* cysticercus antigen in serum samples from NCC patients was found positive in 92% and 87.2% by ELISA and IHA technique, respectively, while in CSF samples, the ELISA and IHA were positive in 88% and 84% patients respectively. The IHA technique was reported to be absolutely specific with CSF samples. Both the techniques have been reported non-specific with the use of serum samples. It is suggested that specific antibody detection in CSF samples may prove useful for the diagnosis [17].

Dot-Blot technique for the serodiagnosis of several parasitic infections has been reported with advantages of easy visual interpretation. With the use of CSE antigen, the evaluation of ELISA and Dot-Blot for the serodiagnosis of NCC in children attending a tertiary care hospital in north India revealed sensitivity, specificity, positive (PPV) and negative predictive values (NPV) and diagnostic efficacy of 89%, 81%, 79%, 90%, and 85% respectively by ELISA and 89%, 73%, 72.5%, 89% and 82% respectively by Dot-Blot. The sensitivity of both ELISA and Dot-Blot techniques was found 100% in children with multiple brain lesions and 87% with a single lesion. It is suggested that for serodiagnosis, application of ELISA with CSE antigen may be better than Dot-Blot technique due to higher specificity and diagnostic efficacy [18]. Further, on comparing the use of CSE and lower molecular mass (LMM) antigens with the same serum samples as used in the previous study [18], the specificity of ELISA and Dot-Blot was 81% and 73% with CSE antigen and 100% and 98% with LMM antigen respectively. No significant difference was observed in sensitivity and specificity with the use of LMM antigen fraction by ELISA and Dot-Blot. It is suggested that Dot-Blot is simpler technique and its application with the LMM antigen may be useful for seroepidemiological surveys [19].

Excretory-Secretory Antigens

ES antigens have been found useful for the serodiagnosis of helminthic infections, yet reports of its application for human NCC are limited. Antibody detection to *T. solium* metacestode ES antigen in CSF samples by ELISA showed 22 positives out of 24 cases with active NCC, and seven patients including six patients with calcified cysts (inactive NCC) and one in a transitional stage were negative. The study highlights that results of ELISA with the use of ES antigen revealed a significant difference between active and inactive NCC [20].

ELISA applied for antibody detection against ES, CSE and LMM antigens revealed a sensitivity of 63.2%, 38.4% and 30.4% with serum samples and 44%, 46.4% and 47.2% with urine samples respectively. The specificity with the use of these antigens was found to be 76.8%, 88% and 85.6% with serum and 65.2%, 66.4% and 58.4% with urine samples respectively. The study suggested that the use of ES antigen for detection of antibody in serum may prove useful for the diagnosis of NCC [21]. The enzyme-linked immunoelectrotransferblot (EITB) assay demonstrated a sensitivity of 85.6% with ES antigen, 80.8% with LMM (10-30 kDa) antigen in serum, and 76.8% and 50.4% in urine samples respectively. The specificity was 64% with ES antigen and 61.6% with LMM antigens in serum and 48% and 33.6% in urine samples respectively. It is thus suggested that the use of EITB assay for antibody detection in serum may be useful for the diagnosis of NCC [22].

Purified Antigenic Fractions

The use of PAF for antibody detection have yielded promising results, although the purification procedures are often complex and require specific equipment and expertise. Different molecular mass antigenic fractions with variable diagnostic efficacy have been reported from different geographical areas (Table). Sephadex G 200 purified antigenic fractions of *T. solium* cysticerci revealed three LMM immunodiagnostic fractions (18, 20 and 24 kDa) with serum samples from NCC patients. The immunoblot analysis indicated 20 kDa fraction to have a high diagnostic efficacy [23]. Further, with the use of PAF in ELISA format, 95% sensitivity and 100% specificity were observed. The PAF 11 fraction is glycoprotein in nature and heat stable at 60°C, a property which may prove useful for epidemiological surveys [24]. The evaluation of seven specific lentil-lectin glycoprotein (LLGP) antigens indicated that p50 band was reactive with sera from all the NCC patients with multiple lesions. It is highlighted that the positivity of EITB assay was 100% for the detection of multi-lesional patients, as compared to ELISA (80%), although lower positivity was reported by both EITB (49%) and ELISA (57%) techniques for the diagnosis of patients with a single small enhancing lesion (SSEL) [25].

T. solium cysticercus antigen purified by polyacrylamide gel-electrophoresis (PAGE) followed by electroelution technique recognised 13, 17 and 26 kDa fractions by immunoblot with sera from NCC patients. ELISA using the purified proteins showed 53%, 88% and 100% specificity with the use of samples from other parasitic infections and syphilis [26]. Further studies in this direction, including samples from patients with CNS tuberculosis are desired to assess the specificity. In another report, the use of *T. solium* cysticercus glycoprotein antigens in ELISA and immunoblot techniques revealed 80% positivity by ELISA and 62% by EITB in patients with SSEL. The specificity was found 100% with the use of samples from patients with CNS tuberculosis. The proteins of ≤18 kDa were reported of significant diagnostic value in patients with SSEL [27]. Bacterially expressed recombinant 10 kDa protein has been reported a reliable marker for the diagnosis of active stage NCC in another study, wherein ELISA was positive in 91.1% of serum and 97.8% of CSF samples. The specificity was high with negligible cross reactions (3.6%) with samples from subjects with other helminthic infections and healthy controls. In contrast, high degree of cross-reactivity was found with the use of crude cyst fluid antigen [28].

T. solium cysticercus fluid purified on the basis of size and ion-exchange chromatography revealed antigenic fraction with cathepsin L-like activity, including 53 and 25 kDa antigenic fractions of diagnostic value. The sensitivity of Western Immunoblot and ELISA was 96% and 98% for patients with multiple lesions, and 78% and 84% for patients with single lesion, while the specificity was 98% and 92.7% respectively with the use of samples from other helminthic infections [29]. Further studies are desired to include samples from patients with clinically similar pathologies for evaluation of specificity. The detection of antibody response to ES antigens has shown promising results for the diagnosis of parasitic diseases. The potential use of *T. solium* metacestode 43 kDa ES antigenic peptide as a diagnostic marker for human NCC has been suggested to yield absolute specificity, while the sensitivity with the use of serum and CSF was 88.23% and 89.28% respectively [30]. The sensitivity of *T. solium* Ag5, cysticercus protein with trypsin-like activity was 96.3%, 75.4% and 39.6% in patients with extraparenchymal cysts, multiple and single cysts respectively; the specificity was 76.7% with the use of other helminthic infected samples and healthy controls [31]. The studies in other CNS pathologies which are clinically similar to NCC may throw further light on specificity.

Table. Purified antigenic fractions with diagnostic efficacy for human neurocysticercosis

Source	Diagnostic utility of antigenic fractions	Study area [Ref.]
Taenia solium cysticerci	Sephadex G 200 PAF revealed three LMM immunodiagnostic fractions (18, 20 and 24 kDa); immunoblot analysis indicated 20 kDa fraction with high diagnostic efficacy; with the use of PAF in ELISA, 95% sensitivity and 100% specificity were observed	North India [23, 24]
	Of seven specific LLGP antigens, p50 band was reactive with 100% serum samples from NCC patients with multiple lesions	North India [25]
	Antigen purified by PAGE followed by electroelution technique recognised 13, 17 and 26 kDa fractions by immunoblot; the specificity of ELISA with the use of 26 kDa was 100%	Brazil [26]
	Cysticercus glycoprotein antigens in ELISA and immunoblot techniques revealed 80% positivity by ELISA and 62% by EITB in patients with SSEL; the proteins of ≤18 kDa were found of significant diagnostic value in patients with SSEL	South India [27]
	Bacterially expressed recombinant 10 kDa protein, used in ELISA showed sensitivity of 91.1% and 97.8% with the use of serum and CSF samples, respectively	Korea [28]
	Cysticercus fluid antigenic fraction, purified by size exclusion and ion-exchange chromatography with cathepsin L-like activity showed 53 and 25 kDa antigenic fractions of diagnostic value	Peru [29]
	The sensitivity for antibody detection to excretory-secretory 43 kDa antigenic peptide was 88.23% and 89.28% with the use of serum and CSF samples respectively, with absolute specificity	South India [30]
	The sensitivity of *T. solium* Ag5, cysticercus protein with trypsin-like activity was 96.36%, 75.44% and 39.62% in patients with extraparenchymal cysts, multiple and single cysts, respectively	Peru [31]
Taenia saginata	*T. saginata* metacestode antigenic fractions obtained by ion-exchange chromatography has shown high diagnostic value; 93.4% sensitivity and 92.6% specificity with DEAE S2, and 88.8% sensitivity and 93.7% specificity with CM S2 fraction	Brazil [32]

Abbreviations: CSF: cerebrospinal fluid; CML: carboxymethyl; DEAE: diethylaminoethyl; ELISA: enzyme-linked immunosorbent assay; EITB: enzyme-linked immunoelectrotransfer blot; LLGP: lentil-lectin glycoprotein; LMM: lower molecular mass; PAGE: polyacrylamide gel electrophoresis; PAF: purified antigenic fractions; SSEL: single space enhancing lesion.

Taenia saginata antigenic fractions obtained by ion-exchange chromatography in diethylaminoethyl sepharose (DEAE) and carboxymethyl sepharose (CM) resins have shown 93.4% sensitivity with DEAE S2 and 88.8% with CM fractions, while the specificity with the use of samples from other parasitic infections was 92.6% with DEAE and 93.7% with CM fraction, thereby indicating high diagnostic value [32]. The samples from patients clinically mimicking NCC, especially tuberculosis, need to be evaluated for assessing specificity for its application in tuberculosis endemic areas.

Recombinant Antigens

The review focused on immunological and molecular diagnosis of cysticercosis indicates that the LLGP-EITB assay is optimal for clinical diagnosis and antigen detection for post-treatment follow-up, while molecular techniques have not been found useful for diagnosis in clinical settings [33]. To address the difficulties in extraction and purification of the native lentil-lectin bound glycoproteins (LLGP) from *T. solium* cysts and the challenges in transferring the purified antigens to other laboratories, recombinant or synthetic antigens were cloned from seven diagnostic glycoproteins of *T. solium* belonging to the three antigenic protein families, i.e., gp50, gp24 and 8 kDa families in the LLGP fraction. Recombinant antigens from these three glycoprotein families (rGP50, rT24H, and TsRS1) were evaluated using an immunoblot assay on serum samples from 249 confirmed NCC-positive patients and were found to have a sensitivity as well as specificity of 99% for diagnosing NCC when all three were used in combination or when rT24H was used alone. The results were comparable to the conventional native LLGP-EITB assay. These developments are encouraging since the recombinant antigen based EITB are easier to replicate across laboratories and the use of such standardized antigens circumvents the dependence on the tedious process of protein extraction from cysts from infected pigs, besides eliminating the lot-to-lot variability [34].

ANTIGEN DETECTION

The studies directed towards antigen detection in serum and body fluids have indicated encouraging results. The cysticercus antigen detection in CSF by ELISA using polyclonal antisera revealed sensitivity of 81.2%, 90% and 95.8% and specificity of 82%, 98% and 100% with the anti-*T. solium* (anti-Tso), anti-*Taenia crassiceps* cysticercal vesicular fluid (anti-Tcra), and anti-Tcra <30 kDA sera respectively. Further, the 14 and 18 kDa peptides were only detected in CSF samples from NCC patients by immunoblotting with anti-Tso and anti-Tcra sera. It is concluded that antigen identification may be considered as an additional marker for the diagnosis and prognosis of NCC [35]. The *T. solium* antigen B fraction is regarded as specific for the diagnosis. An attempt to detect antigen B by ELISA in various fractions of CSF obtained by high performance liquid chromatography (HPLC), revealed positive results in 9 out of 10 CSF samples from NCC patients, yet a positive result was also detected in 9 out of 20 samples from patients with other neurological disorders. False positive results were encountered in 45% patients with CNS tuberculosis. It is suggested that the antigen B detection may be useful for diagnosis in countries free of tuberculosis [36]. The HP 10 antigen detection by ELISA showed similar high sensitivity in serum (84.8%) and CSF (91.3%) in severe NCC patients and was rarely detected in asymptomatic or mild NCC patients. It is concluded that HP 10 detection is useful for the diagnosis of severe NCC patients [37].

Urine and Saliva Samples

Non-invasive samples may be more convenient in terms of sample collection and patient compliance. Specific antigen detection in urine samples of clinically suspected NCC patients by co-agglutination technique revealed positive response in 55.5% clinically suspected NCC patients, 62.5% CT/MRI proven cases and 9% patients with non-cysticercal CNS infections, while all the normal healthy controls were found negative [38]. It is suggested that co-agglutination technique can be used in the field or in a routine clinical laboratory, where facilities for molecular techniques and/or expertise may not be available. Under field conditions in Ecuador and Zambia, the antigen detection revealed 2.85% serum and 12% urine samples positive, while 1.1% samples were positive for both urine and serum Ag by ELISA. The sensitivity of urine Ag ELISA was 86% and no statistically significant difference was found between urine and serum Ag ELISA. The specificity was higher with the use of serum samples as compared to urine. The low specificity with the use of urine samples in this study is suggested to be possibly due to mixed infected and non-infected subjects under study, delayed clearance of cysticercal antigens in the urinary system and/or use of diluted serum and undiluted urine samples [39].

The comparative analysis of Ig subtypes and IgG subclass responses to *T. solium* cysticerci CSE antigen and antigen B in saliva and serum samples showed IgG ELISA sensitivity 71.4% in saliva and 78.9% in serum with CSE antigen, and 65.2% in saliva and 68.2% in serum with antigen B, indicating higher sensitivity with the use of serum samples as compared to saliva. IgG1 ELISA was found to be the most specific (>90%) in both samples [40].

NUCLEIC ACID DETECTION

Polymerase Chain Reaction (PCR)

The approach to detect *T. solium* DNA in the CSF samples has been reported with success. The application of PCR technique revealed *T. solium* DNA in 96.7% samples from NCC patients. It is thus suggested that parasite DNA detection may have significant impact in the NCC diagnosis [41]. The application of real-time PCR technique with the use of CSF samples indicated positivity rate of 83.3% and 100% specificity in patients residing in France. On post-treatment follow-up, the DNA load in CSF either increased or decreased. It is suggested that the increase may be due to the release of parasite cyst contents in the CSF following antiparasitic therapy. The obstruction of the flow of CSF could lead to a false negative PCR in a small number of patients [42]. The comparison of various diagnostic techniques using CSF showed that PCR for DNA detection had the highest sensitivity (95.9%) and variable specificity (80% or 100%) depending upon the controls used, compared to ELISA, EITB and HP10 antigen detection. It is suggested that the results may help in selection of techniques, either singly or in combination depending upon the different requirements [43].

Recently, a highly sensitive quantitative PCR was developed for the diagnosis of subarachnoid and ventricular NCC with encouraging results. A different approach for nucleic acid extraction was used for the circulating cell free DNA of *T. solium* in a magnetic bead-based extraction protocol. The TsolR13 repeat region of *T. solium* DNA was successfully amplified in 81% of 46 plasma samples and 100% of the 36 CSF samples obtained from these patients, thus indicating that DNA based protocols must be encouraged for detection of *T. solium* DNA from CSF samples in extra-parenchymal NCC which is difficult to diagnose by neuroimaging [44]. However, the DNA based assays need more evaluation in patients with parenchymal cysts and with blood samples.

Loop-Mediated Isothermal Amplification (LAMP)

LAMP technique is a unique and novel amplification technique with high specificity and sensitivity and has been reported to have the potential to be used as a point-of care test for the diagnosis of neglected parasitic infections. The advantages reported are based on the ability of LAMP to amplify DNA under isothermal conditions, thus it can be performed using a water/heat bath and eliminates the need of a thermocycler. A positive LAMP test can be visualized with naked eyes without the need of gel electrophoresis [45].

The application of LAMP for diagnosing parasitic infections has revealed certain advantages of this technique over the conventional DNA amplification techniques [46]. For *Taenia* spp., LAMP has been found to be rapid, simple, specific, and sensitive for detection and differentiation of the three species *T. solium*, *T. asiatica* and *T. saginata* with primers targeting the *cox1* gene in stool specimens. A high specificity of LAMP was demonstrated on DNA samples extracted from proglottids and cysticerci of the worms, wherein LAMP could detect one copy of the targeted gene sequence or five eggs of *T. asiatica* and *T. saginata* per gram of the stool specimen, showing a similar sensitivity as that of the PCR methods [47]. Its applicability to rapidly detect *T. solium* at a low cost and infrastructure makes LAMP a valuable tool for potential use in field conditions. Further studies directed towards its application for the diagnosis of NCC may provide additional strategies to the existing knowledge.

Genotype Variation in the Parasite

The analysis of mitochondrial DNA sequences has important considerations in various aspects of NCC [48]. Genotyping helps to understand the evolutionary aspects of the emergence of two different phylogenies of the parasite, i.e., the Asian and the African/American strains of *T. solium*. The genomic studies may also throw light on the differences in the manifestations of NCC with or without subcutaneous cysticercosis in the two strains. The genetic signatures of the parasite have been found as a useful marker to confirm the diagnosis of NCC in clinically suspected patients. The diagnosis in a patient with solitary cyst was confirmed by mitochondrial DNA analysis, revealing Asian genotype of *T. solium* [49]. In a Mongolian traveller who manifested epilepsy due to NCC after visiting to China and India, mitochondrial *cox1* gene analysis revealed that he was affected by the Indian haplotype of *T. solium* and not the Chinese haplotype [50].

Thus, genotyping may aid in molecular tracking of the transmission and epidemiology of the disease.

PROTEOMICS

Since proteins are the main catalysts and molecular fulcrums of biological tissues, the studies focusing on the proteome may be able to provide relevant clues for the clinical management of patients. It is suggested that the next generation of assays for the diagnosis of infectious diseases will utilize proteomic techniques. The reported applications of advanced proteomics in parasitic infections, reviewed earlier, highlighted the advantages/disadvantages of different techniques, including 2D-gel electrophoresis, MALDI-TOF and SELDI-TOF [46]. The most useful top-down strategy is 2-dimensional electrophoresis (2D-PAGE). Analysis of LMM (10-30 kDa) antigen fraction of *T. solium* metacestode by 2D-PAGE followed by EITB assay revealed that out of seropositive and seronegative samples from NCC patients, 100% and 60% respectively, were reactive with 2D fraction antigens; the specificity was 92%. It is suggested that 2D-PAGE EITB may serve useful purpose for the diagnosis in endemic areas [51].

CELL MEDIATED TECHNIQUES

Cell mediated immune responses are well documented in human NCC [52]. The application of cyst fluid antigen-based lymphocyte transformation (LTT) assay for the diagnosis of NCC indicated a sensitivity of 93.8% and a specificity of 96.2%. It is suggested that LTT can be used as an immunodiagnostic tool for NCC after validation in endemic areas [53].

CLINICAL CONSIDERATIONS

Presentation vs. Diagnosis

Clinical manifestations of human NCC seem to be variable in different geographical locations. In Asian countries, the majority of the patients host a single enhancing lesion while in Latin America, the most common type of clinical presentation is a few viable cysts without inflammation. The EITB assay usually performs better than ELISA, however, its sensitivity in patients with one degenerating lesion is low [4]. The patients with only calcified cysts, either single or multiple, are also less likely to have an EITB positivity [54]. In addition, the detection of specific antibody response in body fluids is dependent on the developmental stage of the metacestode and type of the technique used [55].

Single vs. Multiple Lesions

Review reports have consistently concluded the difference in antibody response in patients with single and multiple lesions, depending upon the technique used. The positivity of EITB assay was found to be 100% for the diagnosis of multi-lesional NCC as compared to ELISA (80%), while lower positivity was reported in patients with SSEL by both EITB (49%) and ELISA (57%) [25]. A discordance between the results of EITB and ELISA reports and lower sensitivity of EITB for the detection of SSEL may lead to its limited use in endemic areas with predominance of SSEL patients as compared to multi-lesional patients. It is suggested that low sensitivity of EITB in SSEL patients may be due to limited/insufficient immune response provided by a single cyst.

Childhood NCC

Neurocysticercosis in children is a well-documented clinical entity. The extraparenchymal NCC is rare in children and has poor prognosis, while the prognosis in patients with SSEL is good. The positive antibody response has been reported in only 17 to 25% children with NCC and is of low diagnostic value in patients with SSEL. The scheme of diagnostic evaluation in children with suspected parenchymal and extraparenchymal NCC is detailed out in a review which also suggests that based on neuroimaging reports and serodiagnostic techniques, the diagnosis of NCC can be confirmed [56]. The clinical presentation in children may also vary in different endemic countries, thus it is desired to define the diagnostic criteria based on the specific area.

Post-Treatment Follow-Up

Serodiagnostic techniques principally have a screening and/or diagnostic role. A positive antigen detection indicates presence of live parasites and is thus a useful marker for direct therapeutic decisions and/or monitoring. The HP 10 antigen levels dropped in serum and CSF of patients after treatment, in which parasites disappeared or became calcified, while the levels continued to be elevated in patients in whom parasites remained vesicular which suggests that HP 10 antigen detection may be a useful modality for the follow-up of NCC cases [37]. On comparing serological response against the CSE, ES and LMM fractions by ELISA, there was no significant difference in the antibody response in pre- and post-treatment urine samples with the use of the three type of antigens. The study suggests that detection of antibody response to ES antigens in serum samples may help to assess the therapeutic response [57]. In post-treatment saliva, IgG4 subclass has been reported to be undetectable, thus indicating that IgG4 ELISA detection in saliva samples may serve as a better marker for NCC follow-up [40]. Of the nucleic acid amplification assays, TsolR13 qPCR has been suggested to have a potential to be used as a test of cure of the disease when CSF samples are used for follow-up post-treatment [44]. Further studies focused on assessing the technique(s) for sequential long time periods following treatment may throw light to define the best possible marker to evaluate treatment success and to formulate strategies for follow-up.

CONCLUSION

The diagnosis of NCC is challenging. The radio-imaging reports may mimic other diseases of the nervous system, posing problems in confirmation of diagnosis. Moreover, these techniques may not serve a use for epidemiological studies. The main issue of serodiagnostic techniques is their sensitivity, which depends mainly upon the technique(s) utilized and presence of single vs. multiple lesions. The lack of specificity has been attributed to the type of diagnostic antigen used and/or presence of other clinically similar infections in an endemic area, posing problems of cross-reactivity. Molecular techniques providing higher sensitivity are not usually applied in endemic areas due to non-availability of facilities and expertise at the diagnostic centres in the developing countries and are expensive.

There are conflicting reports from different geographical areas and variability in the conclusions drawn from myriad diagnostic studies. It is pertinent to note that such a variation between the reports may be due to difference in the type of population under study (asymptomatic vs. symptomatic), clinical manifestations in infected subjects in terms of the number of lesions, stage of infection, type of cysts and type of control subjects (normal healthy subjects, other disease control, or clinically similar pathologies). In addition, the methodology adopted for interpretation of the data may contribute to the variance in the different reports. In a few reports, the number of confirmed NCC patients based on radio-imaging techniques have been taken into account for calculation of sensitivity of the serodiagnostic techniques, while in others the numbers of clinically suspected NCC patients have been taken into consideration.

Similarly, for specificity, in a few reports only healthy controls have been considered, in some others the disease controls were also included, and in very few, patients with clinically similar pathologies have also been taken into account. Thus, it may not be a straight forward task to compare the reports from different geographical areas in terms of sensitivity, specificity and diagnostic efficacy. The sensitivity based on a number of confirmed patients with more than one technique may provide some justifiable conclusions.

Recent, a possibility of *Taenia asiatica* leading to cysticercosis is reported based mainly on its liver tropism in pigs, the intermediate host both for *T. solium* and *T. asiatica*. With this possibility, the application of conventional morphological and serological approach may pose problems in the specific diagnosis, as tests targeting *T. solium* cysticerci may cross react with *T. asiatica*. The molecular techniques can, however, differentiate the two *Taenia* species, thereby suggesting the application of molecular techniques for specific diagnosis of *T. asiatica/T. solium* cysticercosis etiology [58].

The main challenges for the diagnosis of NCC are to identify the best criteria for confirmatory diagnosis, in view of different clinical presentation in NCC patients and endemicity of other clinically similar pathologies in different geographical areas. It is suggested that the ideal diagnostic test(s) should be non-expensive, simple to perform, point-of-care technique, with good sensitivity and specificity, both for use in patients with solitary and multiple cyst lesions [7]. Further, a single criterion for diagnosis and/or similar criteria for all the endemic areas may not provide the desired results. The criteria need to be defined, should be specific for the endemic area, keeping in view the base line factors, in terms of clinical presentation and other similar endemic pathologies. A combination of techniques for serodiagnosis with the use of sensitive technique and specific antigens may prove useful in settings with limited resources for molecular techniques.

REFERENCES

[1] Mahajan, R. C. (1982). Geographical distribution of human cysticercosis. In Flisser, A., Willms, K., Laclette, C., Larrolde, C., Ridaura, C. and Beltran, F. (Eds.), *Cysticercosis - Present State of Knowledge and Perspectives* (pp. 39-46). New York: Academic Press.

[2] Carpio, A. (2002). Neurocysticercosis: an update. *The Lancet Infectious Diseases*, 2: 751-62.

[3] Ito, A., Nakao, M. and Wandra, T. (2003). Human Taeniasis and cysticercosis in Asia. *The Lancet*, 362: 1918-1920.

[4] García, H. H., Gonzalez, A. E., Evans, C. A. W., Gilman, R. H., Cysticercosis Working Group in Peru (2003). *Taenia solium* cysticercosis. *The Lancet*, 362: 547-556.

[5] World Health Organization (2012). Accelerating work to overcome the global impact of neglected tropical diseases: a roadmap for implementation. Geneva: World Health Organization.

[6] Malla, N. (2005). Human cysticercosis in India: a zoonotic disease of public health importance. In Tandon, V. and Dhawan, B. N. (Eds.), *Infectious diseases of domestic animals and zoonosis in India* (pp. 197-204). *Proceedings of the National Academy of Sciences*, India.

[7] Mewara, A., Goyal, K. and Sehgal, R. (2013). Neurocysticercosis: A Disease of Neglect. *Indian Journal of Tropical Parasitology*, 3: 106-113.

[8] Rajshekhar, V. (2016). Neurocysticercosis: Diagnostic problems & current therapeutic strategies. *Indian Journal of Medical Research*, 144: 319-326.

[9] Ahmad, R., Khan, T., Ahmad, B., Misra, A., and Balapure, A. K. (2017). Neurocysticercosis: a review on status in India, management, and current therapeutic interventions. *Parasitology Research*, 116: 21-33.

[10] Grover, R., Mahajan, R. C. and Malla, N. (2002). Increasing seroprevalence to cysticercus cellulosae antigen in clinically

suspected Neurocysticercosis children. *Indian Journal of Pathology and Microbiology*, 45: 307-311.

[11] Del Brutto, O. H., Rajshekhar, V., White, A. C. Jr., Tsang, V. C., Nash. T. E., Takayanagui, O. M., et al. (2001). Proposed diagnostic criteria for neurocysticercosis. *Neurology*, 57: 177-183.

[12] Garg, R. K. (2004). Diagnostic criteria for neurocysticercosis: some modifications are needed for Indian patients. *Neurology India*, 52: 171-177.

[13] Del Brutto, O. H. (2012). Diagnostic criteria for neurocysticercosis, revisited. *Pathogens and Global Health*, 106: 299-304.

[14] Del Brutto, O. H., Nash, T. E., White, A. C. Jr., Rajshekhar, V., Wilkins, P. P., Singh, G. et al. (2017). Revised diagnostic criteria for neurocysticercosis. *Journal of the Neurological Sciences*, 372: 202-210.

[15] White, A. C. Jr., Coyle, C. M., Rajshekhar, V., Singh, G., Hauser, W. A., Mohanty, A., et al. (2018) Diagnosis and Treatment of Neurocysticercosis: 2017 Clinical Practice Guidelines by the Infectious Diseases Society of America (IDSA) and the American Society of Tropical Medicine and Hygiene (ASTMH). *Clinical Infectious Diseases*, 66: e49-e75.

[16] White, A. C. Jr. (1997). Neurocysticercosis: A major cause of neurological disease worldwide. *Clinical Infectious Diseases*, 24: 101-113.

[17] Malla, N., Kaur, M., Kaur, U., Ganguly, N. K., and Mahajan, R. C. (1992). Evaluation of Enzyme Linked Immunosorbent Assay for the detection of anticysticercus antibodies in cerebrospinal fluid from patients with neurocysticercosis. *Journal of hygiene, epidemiology, microbiology, and immunology*, 36: 181-190.

[18] Mandal, J., Singhi, P. D., Khandelwal, N. and Malla, N. (2006). Evaluation of ELISA and dot blots for the serodiagnosis of neurocysticercosis, in children found to have single or multiple enhancing lesions in computerized tomographic scans of the brain. *Annals of Tropical Medicine & Parasitology*, 100: 39-48.

[19] Mandal, J., Singhi, P. D., Khandelwal, N. and Malla, N. (2008). Evaluation of lower molecular mass (20-24 kDa) *Taenia solium* cysticercus antigen fraction by ELISA and dot blot for the serodiagnosis of neurocysticercosis in children. *Parasitology Research*, 102: 1097-1101.

[20] Molinari, J. L., Mendoza, E. G., de la Garza, Y., Ramírez, J. A., Sotelo, K. and Tato, J. (2002). Discrimination between active and inactive neurocysticercosis by metacestode excretory/secretory antigens of *Taenia solium* in an enzyme-linked immunosorbent assay. *The American Journal of Tropical Medicine and Hygiene*, 66: 777-781.

[21] Atluri, S. R. V., Singhi, P., Khandelwal, N. and Malla, N. (2009). Neurocysticercosis immunodiagnosis using *Taenia solium* cysticerci crude soluble extract, excretory secretory and lower molecular mass antigens in serum and urine samples of Indian children. *Acta Tropica*, 110: 22-27.

[22] Atluri, S. R. V., Singhi, P., Khandelwal, N. and Malla, N. (2009). Evaluation of excretory secretory and 10-30 kDa antigens of *Taenia solium* Cysticerci by EITB assay for the diagnosis of neurocysticercosis. *Parasite Immunology,* 31: 151-155.

[23] Kaur, M., Ganguly. N, K., Mahajan, R. C. and Malla, N. (1995). Identification of antigenic fractions of *Cysticercus cellulosae* by western blotting in the serodiagnosis of human neurocysticercosis: before and after treatment. *Immunology and Infectious Diseases,* 5: 67-72.

[24] Kaur, M., Goyal, R., Ganguly, N. K., Mahajan, R. C. and Malla, N. (1996). Evaluation and characterization of purified antigenic fraction-II of *Cysticercus cellulosae* by enzyme-linked immunosorbent assay for the diagnosis of neurocysticercosis: before and after treatment. *Immunology and Infectious Diseases,* 6: 25-29.

[25] Singh, G., Kaushal, V., Ram, S., Kaushal, R. K., Dhanuka, A. K. and Khurana, S. (1999). Cysticercus immunoblot assay in patients with single, small enhancing lesions and multilesional neurocysticercosis. *Journal of the Association of Physicians of India,* 47: 476-479.

[26] Ev, L. V., Maia, A. A. M., Pianetti, G. and Nascimento, E. (1999). Immunological evaluation of a 26-kDa antigen from *Taenia solium* larvae for specific immunodiagnosis of human neurocysticercosis. *Parasitology Research*, 85: 98-102.

[27] Prabhakaran, V., Rajshekhar, V., Murrell, K. D. and Oommen, A. (2004). *Taenia solium* metacestode glycoproteins as diagnostic antigens for solitary cysticercus granuloma in Indian patients. *Transactions of the Royal Society of Tropical Medicine and Hygiene*, 98: 478-484.

[28] Lee, E. G., Lee, M. Y., Chung, J. Y., Je, E. Y., Bae, Y. A., Na, B. K., et al. (2005). Feasibility of baculovirus-expressed recombinant 10-kDa antigen in the serodiagnosis of *Taenia solium* neurocysticercosis. *Transactions of the Royal Society of Tropical Medicine and Hygiene*, 99: 919-926.

[29] Zimic, M., Pajuelo, M., Rueda, D., López, C., Arana, Y., Castillo, Y., et al. (2009). Utility of a Protein Fraction with Cathepsin L-Like Activity Purified from Cysticercus Fluid of *Taenia solium* in the Diagnosis of Human Cysticercosis. *The American Journal of Tropical Medicine and Hygiene*, 80: 964-970.

[30] Sahu, P. C., Parija, S. C. and Jayachandran, S. (2010). Antibody specific to 43kDa excretory-secretory antigenic peptide of *Taenia solium* metacestode as a potential diagnostic marker in human neurocysticercosis. *Acta Tropica*, 115: 257-261.

[31] Rueda, A., Sifuentes, C., Gilman, R. H., Gutiérrez, A. H., Piña, R., Chile, N., et al. (2011). TsAg5, a *Taenia solium* cysticercus protein with a marginal trypsin-like activity in the diagnosis of human neurocysticercosis. *Molecular and Biochemical Parasitology*, 180: 115-119.

[32] da Silva Nunes, D., Gonzaga, H. T., da Silva Ribeiro, V., da Cunha, Jr. J. P. and Costa-Cruz, J. M. (2013). *Taenia saginata* metacestode antigenic fractions obtained by ion-exchange chromatography: potential source of immunodominant markers applicable in the immunodiagnosis of human neurocysticercosis. *Diagnostic Microbiology and Infectious Disease*, 76: 36-41.

[33] Rodriguez, S., Wilkins, P. and Dorny, P. (2012). Immunological and molecular diagnosis of cysticercosis. *Pathogens and Global Health,* 106: 286-298.
[34] Noh, J., Rodriguez, S., Lee, Y. M., Handali, S., Gonzalez, A. E., Gilman, R. H., et al. (2014) Recombinant protein- and synthetic peptide-based immunoblot test for diagnosis of neurocysticercosis. *Journal of Clinical Microbiology,* 52: 1429-1434.
[35] Pardini, A. X., Vaz, A. J., Machado, L. D. R. and Livramento, J. A. (2001). Cysticercus antigens in cerebrospinal fluid samples from patients with neurocysticercosis. *Journal of Clinical Microbiology,* 39: 3368-3372.
[36] Das, S., Mahajan, R. C., Ganguly, N. K., Sawhney, I. M. S., Dhawan, V. and Malla, N. (2002). Detection of antigen B of *Cysticercus cellulosae* in cerebrospinal fluid for the diagnosis of human neurocysticercosis. *Tropical Medicine & International Health,* 7: 53-58.
[37] Fleury, A., Hernández, M., Avila, M., Cárdenas, G., Bobes, R. J., Huerta, M., et al. (2007) Detection of HP10 antigen in serum for diagnosis and follow-up of subarachnoidal and intraventricular human neurocysticercosis. *Journal of Neurology, Neurosurgery and Psychiatry,* 78: 970-974.
[38] Parija, M., Biswas, R., Harish, B. N. and Parija, S. C. (2004). Detection of specific cysticercus antigen in the urine for diagnosis of neurocysticercosis. *Acta Tropica,* 92: 253-260.
[39] Mwape, K. E., Praet, N., Benitez-Ortiz, W., Muma, J. B., Zulu, G. and Celi-Erazo, M. (2011). Field evaluation of urine antigen detection for diagnosis of *Taenia solium* cysticercosis. *Transactions of the Royal Society of Tropical Medicine and Hygiene,* 105: 574-578.
[40] Malla, N., Kaur, R., Ganguly, N. K., Sawhney, I. M. and Mahajan, R. C. (2005). Utility of specific IgG4 response in saliva and serum samples for the diagnosis and follow up of human neurocysticercosis. *Nepal Medical College Journal,* 7: 1-9.

[41] Almeida, C. R., Ojopi, E. P., Nunes, C. M., Machado, L. R., Takayanagui, O. M., Livramento, J. A., et al. (2006). *Taenia solium* DNA is present in the cerebrospinal fluid of neurocysticercosis patients and can be used for diagnosis. *European Archives of Psychiatry and Clinical Neuroscience,* 256: 307-310.

[42] Yera, H., Dupont, D., Houze, S., M'Rad, M. B., Pilleux, F., Sulahian, A., et al. (2011). Confirmation and Follow-Up of Neurocysticercosis by Real-Time PCR in Cerebrospinal Fluid Samples of Patients Living in France. *Journal of Clinical Microbiology,* 49: 4338-4340.

[43] Michelet, L., Fleury, A., Sciutto, E., Kendjo, E., Fragoso, G., Paris, L., et al. (2011). Human neurocysticercosis: comparison of different diagnostic tests using cerebrospinal fluid. *Journal of Clinical Microbiology,* 49: 195-200.

[44] O'Connell, E. M., Harrison, S., Dahlstrom, E., Nash, T. and Nutman, T. B. (2020). A Novel, Highly Sensitive Quantitative Polymerase Chain Reaction Assay for the Diagnosis of Subarachnoid and Ventricular Neurocysticercosis and for Assessing Responses to Treatment. *Clinical Infectious Diseases*, 70: 1875-1881.

[45] Avendaño, C. and Patarroyo, M. A. (2020). Loop-Mediated Isothermal Amplification as Point-of-Care Diagnosis for Neglected Parasitic Infections. *The International Journal of Molecular Sciences,* 21: 7981.

[46] Ndao, M. (2009). Diagnosis of Parasitic Diseases: Old and New approaches. *Interdisciplinary Perspectives on Infectious Diseases,* 278246.

[47] Nkouawa, A., Sako, Y., Nakao, M., Nakaya, K. and Ito, A. (2009). Loop-mediated isothermal amplification method for differentiation and rapid detection of *Taenia* species. *Journal of Clinical Microbiology,* 47: 168-174.

[48] Ito, A., Yamasaki, H., Nakao, M., Sako, Y., Okamoto, M., Sato, M. O., et al. (2003). Multiple genotypes of *Taenia solium* - ramifications for diagnosis, treatment and control. *Acta Tropica,* 87: 95-101.

[49] Yamasaki, H., Matsunaga, S., Yamamura, K., Chang, C., Kawamura, S., Sako, Y., et al. (2004). Solitary neurocysticercosis case caused by Asian genotype of *Taenia solium* confirmed by mitochondrial DNA analysis. *Journal of Clinical Microbiology,* 42: 3891-3893.

[50] Davaasuren, A., Davaajav, A., Ukhnaa, B., Purvee, A., Unurkhaan, S., Luvsan, A., et al. (2017). Neurocysticercosis: A case study of a Mongolian traveler who visited China and India with an updated review in Asia. *Travel Medicine and Infectious Disease,* 20: 31-36.

[51] Atluri, V. S. R., Singhi, P. D., Khandelwal, N. and Malla, N. (2011). 2D-PAGE analysis of *Taenia solium* metacestode 10-30 kDa antigens for the serodiagnosis of neurocysticercosis in children. *Acta Tropica,* 118: 165-169.

[52] Grewal, J. S., Kaur, S., Bhatti, G., Sawhney, I. M., Ganguly, N. K., Mahajan, R. C., et al. (2000). Cellular immune responses in human neurocysticercosis. *Parasitology Research,* 86: 500-503.

[53] Prasad, A., Prasad, K. N., Yadav, A., Gupta, R. K., Pradhan, S., Jha, S., et al. (2008). Lymphocyte transformation test: a new method for diagnosis of neurocysticercosis. *Diagnostic Microbiology and Infectious Disease,* 61: 198-202.

[54] Wilson, M., Bryan, R. T., Fried, J. A., Ware, D. A., Schantz, P. M., et al. (1991). Clinical Evaluation of the Cysticercosis Enzyme-Linked Immunoelectrotransfer Blot in Patients with Neurocysticercosis. *The Journal of Infectious Diseases,* 164: 1007-1009.

[55] Lopez, J. A., Garcia, E., Cortes, I. M., Sotelo, J., Tato, P. and Molinari, J. L. (2004). Neurocysticercosis: relationship between the developmental stage of metacestode present and the titre of specific IgG in the cerebrospinal fluid. *Annals of Tropical Medicine and Parasitology,* 98: 569-579.

[56] Singhi, P. and Singhi, S. (2004). Neurocysticercosis in Children. *Journal of Child Neurology,* 19: 482-492.

[57] Atluri, V. S. R., Gogulamudi, V. R., Singhi, P., Khandelwal, N., Parasa, L. S. and Malla, N. (2014). Pediatric Neurocysticercosis: Usefulness of Antibody Response in Cysticidal Treatment Follow-Up. *BioMed Research International,* ID 904046.

[58] Galán-Puchades, M. T. and Fuentes, M. V. (2013). *Taenia asiatica*: the most neglected human *Taenia* and the possibility of cysticercosis. *The Korean Journal of Parasitology*, 51: 51-54.

INDEX

A

antibody, vii, x, 49, 50, 51, 53, 54, 74, 75, 76, 77, 78, 79, 80, 81, 88, 89
antiepileptic drugs, 8, 34
antigen, vii, ix, 22, 51, 55, 74, 75, 76, 77, 78, 79, 82, 83, 84, 85, 87, 89, 90, 92, 94, 95, 96
antigen B, 52, 83, 84, 96
antigenicity, 50
antioxidant, 39
arachnoiditis, 9
asymptomatic, x, 5, 12, 74, 75, 83, 90

B

bioavailability, 11, 13, 28
bioinformatics, 41
biological processes, 22
biomarkers, 55, 57
biopsy, 45, 75
biosynthesis, 11
biotic, 5

blood, 6, 10, 12, 14, 19, 22, 26, 57, 58, 59, 60, 85
blood circulation, 6
blood-brain barrier, 10, 26
body fluid, vii, ix, 51, 61, 74, 76, 83, 88
brain, 6, 7, 9, 10, 13, 14, 19, 38, 45, 46, 47, 48, 51, 54, 59, 60, 75, 78, 93
breast cancer, 18, 30

C

calcification, 10, 21, 76
calcifications, 13, 45, 76
calcium, 11, 20, 24, 33, 36
cancer, 21, 27
cancer therapy, 27
carbamazepine, 9, 11, 14, 28
cell division, 12, 23
cell signaling, 27
central nervous system, vii, ix, 1, 2, 31, 45, 73, 75
cerebral edema, 8, 15
cerebrospinal fluid, 7, 14, 33, 35, 81, 93, 96, 97, 98
challenges, 25, 31, 35, 36, 55, 82, 91

Index

chemotherapy, 21
children, 3, 29, 75, 77, 89, 93, 94, 98
chromatography, 80, 81, 82, 95
clinical diagnosis, ix, 73, 75, 77, 82
clinical presentation, vii, viii, ix, 44, 45, 74, 76, 88, 89, 91
clinical syndrome, 75
computed tomography, 47, 65
cost, 3, 11, 16, 48, 50, 55, 59, 61, 86
crude soluble extract, x, 62, 74, 77, 94
cyst, 3, 5, 6, 8, 13, 21, 22, 47, 48, 49, 50, 51, 52, 53, 57, 58, 59, 76, 80, 85, 86, 87, 88, 91
cysticercosis, ix, 2, 5, 8, 9, 10, 13, 22, 23, 25, 27, 30, 32, 33, 34, 35, 36, 38, 39, 40, 41, 45, 52, 53, 58, 64, 65, 66, 68, 70, 71, 73, 74, 82, 86, 91, 92, 95, 96, 98, 99
cysticercus, 6, 37, 41, 47, 52, 69, 77, 79, 80, 81, 83, 92, 94, 95, 96
cytochrome, 58

D

data mining, 22, 28, 41
detection, vii, ix, 25, 47, 48, 49, 50, 51, 52, 54, 56, 58, 59, 74, 76, 77, 78, 79, 80, 81, 82, 83, 84, 85, 86, 88, 89, 93, 96, 97
developed countries, ix, 73, 74
developing economies, vii, 1
developmental process, 20
diagnosis, v, viii, ix, 25, 35, 36, 37, 40, 41, 43, 44, 45, 46, 47, 48, 49, 50, 52, 55, 57, 58, 59, 60, 61, 62, 64, 65, 66, 67, 68, 69, 70, 71, 73, 74, 75, 76, 77, 78, 79, 80, 82, 83, 84, 85, 86, 87, 88, 89, 90, 91, 93, 94, 95, 96, 97, 98
diagnostic criteria, vii, x, 45, 46, 74, 76, 89, 93
diseases, viii, ix, 3, 20, 27, 28, 29, 44, 58, 73, 87, 90, 92

DNA, 55, 56, 57, 59, 61, 64, 66, 68, 84, 85, 86, 97
DNA polymerase, 57, 59
drug design, 22, 24
drug discovery, 15
drug reactions, 14
drug targets, viii, 2, 3, 15, 19, 20, 23, 29, 37
drug therapy, 28, 30, 45
drug toxicity, 12
drug-induced hepatitis, 29
drugs, viii, 2, 3, 7, 8, 10, 12, 14, 15, 16, 21, 22, 26, 31, 36, 37, 38, 39

E

EITB, 48, 49, 50, 51, 55, 64, 78, 79, 80, 81, 82, 85, 87, 88, 94
electrophoresis, 50, 79, 81, 85, 87
ELISA, x, 51, 52, 53, 54, 62, 63, 66, 69, 70, 74, 77, 78, 79, 80, 81, 83, 84, 85, 88, 89, 93, 94
ELISA method, 53
encephalitis, 6, 7, 8, 9, 13
enzyme, 14, 16, 52, 53, 64, 78, 81, 94
enzyme-linked immunoelectrotransferblot, 78
enzyme-linked immunosorbent assay, 64, 77, 81, 94
enzymes, 19, 24, 25, 39
epidemiology, 33, 40, 87, 93
epilepsy, vii, viii, 1, 2, 5, 9, 13, 14, 15, 28, 29, 39, 40, 43, 45, 63, 64, 68, 69, 70, 75, 86
epileptogenesis, 21
equipment, 79
estrogen, 18, 22, 28
excretory-secretory, 77, 78, 81, 95
expertise, 55, 79, 84, 90
exposure, 11, 45, 46, 53
extraparenchymal cyst, 80, 81

F

false negative, 85
false positive, 49, 52, 54, 55, 59
fluid, 6, 22, 41, 48, 51, 52, 53, 58, 80, 81, 83, 87
food, viii, 8, 36, 44
formation, 12, 21, 47, 48
funduscopic examination, 45

G

gastrointestinal tract, 4
genome, 22, 23, 28, 55, 76
genotyping, vii, ix, 74, 87
glucose, 12, 20, 23, 29, 33, 39
glycogen, 12, 21, 39, 40
glycoproteins, 50, 69, 82, 95
guidelines, vii, ix, 9, 13, 32, 45, 46, 76

H

headache, viii, 6, 9, 11, 12, 44
health, vii, viii, 1, 20, 44, 51, 61, 75, 76
high performance liquid chromatography, 83
host, ix, 3, 4, 5, 6, 9, 11, 12, 19, 20, 21, 23, 24, 25, 26, 27, 31, 44, 47, 60, 75, 88, 91
human, viii, x, 2, 4, 20, 26, 30, 31, 35, 37, 41, 44, 53, 56, 58, 60, 64, 74, 76, 78, 80, 81, 87, 88, 92, 94, 95, 96, 98, 99
hydatid disease, 28
hydrocephalus, ix, 7, 13, 44, 45, 48, 53
hygiene, viii, 44, 93

I

immune response, 3, 7, 13, 22, 36, 61, 64, 75, 87, 88, 98
immune system, 9, 11, 12, 25
immunoglobulin, 54
immunostimulatory, 26
in vitro, 16, 18, 23, 31, 39
infection, viii, ix, 4, 7, 17, 25, 26, 34, 36, 40, 44, 45, 47, 49, 52, 53, 54, 58, 60, 73, 74, 90
insulin, 23, 29, 31, 32, 33, 39, 40, 41
insulin signaling, 23, 32
intracranial pressure, ix, 6, 8, 44, 76
ion-exchange, 80, 81, 82, 95

L

larval development, 32, 47
lentil-lectin bound glycoproteins, 82
lesions, 6, 9, 10, 12, 13, 15, 26, 45, 46, 47, 48, 49, 60, 76, 78, 79, 80, 81, 88, 90, 91, 93, 94
liver function tests, 12
loop-mediated isothermal amplification, 68, 85, 97

M

magnetic resonance imaging, 47
management, viii, 2, 7, 9, 12, 26, 27, 35, 36, 37, 39, 40, 54, 61, 76, 87, 92
membrane permeability, 11
meningitis, ix, 44, 53
meta-analysis, 30, 34
metabolism, 10, 16, 19, 21, 22, 23, 39
mitochondrial DNA, 86, 98
modifications, x, 74, 76, 93
molecular mass, x, 74, 78, 79, 81, 94
MRI, 10, 13, 45, 46, 47, 49, 60, 61, 68, 76, 83
muscle contraction, 20

N

neglected tropical disease, ix, 3, 28, 63, 64, 68, 73, 74, 92
nervous system, 3, 30, 90
neurocysticercosis, v, vii, viii, ix, 1, 2, 8, 26, 27, 28, 29, 30, 31, 32, 33, 34, 35, 36, 37, 38, 39, 40, 43, 45, 46, 61, 62, 63, 64, 65, 66, 67, 68, 69, 70, 71, 73, 74, 81, 89, 92, 93, 94, 95, 96, 97, 98, 99
neuroimaging, ix, 10, 15, 30, 45, 46, 47, 51, 76, 85, 89
neurological disease, 93
nucleic acid, vii, ix, 55, 57, 59, 74, 76, 85, 89

P

parasite, viii, 3, 4, 6, 11, 12, 13, 18, 19, 20, 21, 24, 25, 26, 27, 28, 31, 33, 35, 44, 45, 46, 48, 53, 54, 74, 76, 84, 86
parasitic diseases, 80
parasitic infection, vii, 1, 2, 18, 77, 79, 82, 85, 87
parenchyma, vii, 1, 2, 5, 6
peptide, 80, 81, 95, 96
plasma membrane, 36
polymerase, 59
polymerase chain reaction, 55, 68, 84, 97
population, viii, 2, 44, 61, 75, 90
protein synthesis, 18, 23, 39
proteins, 18, 19, 21, 25, 27, 49, 50, 54, 60, 79, 81, 87
proteomics, x, 27, 74, 87
public health, vii, ix, 1, 73, 74, 92
purification, 50, 51, 79, 82
purified antigenic fractions, 77, 79, 81

R

reactivity, 48, 49, 50, 52, 53, 55, 80, 90
receptors, viii, 2, 23, 24, 25, 35, 40, 41
recombinant antigens, 66, 77, 82
recommendations, iv, 13, 28, 35, 36
reproduction, 3, 18, 22, 23, 31, 39
response, x, 5, 6, 9, 11, 15, 20, 47, 54, 74, 77, 80, 83, 88, 89, 96
rural population, 52, 61

S

scolex, 5, 6, 23, 31, 45, 46, 47, 51
seizure, viii, 3, 9, 31, 37, 44, 49, 58
sensitivity, ix, 48, 49, 50, 51, 53, 54, 55, 56, 57, 58, 59, 74, 77, 78, 79, 80, 81, 82, 83, 84, 85, 87, 88, 90, 91
serum, 10, 14, 46, 49, 50, 52, 53, 54, 57, 77, 78, 79, 80, 81, 82, 83, 84, 89, 94, 96
species, 16, 18, 20, 22, 24, 25, 39, 55, 58, 86, 91, 97
structural variation, 19
subcutaneous tissue, 2
surgical intervention, 3, 15
surgical removal, 8
symptoms, 3, 6, 75

T

T. asiatica, 56, 58, 86, 91
T. saginata, 5, 16, 17, 18, 53, 56, 58, 81, 86
Taenia, vii, ix, 1, 2, 4, 5, 6, 18, 20, 22, 23, 24, 25, 27, 30, 31, 32, 33, 34, 35, 36, 37, 38, 39, 40, 41, 55, 58, 59, 61, 62, 64, 65, 66, 67, 68, 69, 71, 73, 74, 81, 82, 83, 86, 91, 92, 94, 95, 96, 97, 98, 99
Taenia solium, vii, ix, 1, 2, 27, 30, 31, 33, 34, 36, 37, 39, 40, 41, 59, 61, 62, 64, 65,

66, 67, 68, 69, 71, 73, 74, 81, 92, 94, 95, 96, 97, 98
Taeniasis, ix, 40, 56, 64, 69, 73, 74, 92
tapeworm, vii, 1, 2, 4, 5, 17, 18, 34, 39
target, 20, 21, 23, 24, 25, 26, 35, 55, 56, 58, 59
techniques, vii, ix, 3, 47, 51, 74, 75, 76, 77, 78, 79, 80, 81, 82, 84, 85, 86, 87, 89, 90, 91
therapeutic interventions, 27, 92
therapeutic targets, 21, 27
therapeutics, viii, 2, 3, 26, 28, 38
therapy, viii, 2, 3, 7, 8, 9, 13, 17, 19, 25, 27, 29, 30, 32, 34, 38, 40, 85
treatment, vii, viii, x, 2, 3, 7, 8, 9, 10, 11, 12, 13, 14, 15, 16, 18, 19, 22, 23, 25, 26, 28, 29, 31, 32, 34, 35, 36, 37, 38, 40, 74, 82, 84, 89, 94, 97
trypsin, 22, 37, 80, 81, 95
tuberculosis, 75, 79, 82, 83

U

urine, 54, 78, 83, 89, 94, 96

V

vaccine, 21, 23, 25, 27
ventriculoperitoneal shunt, 8, 14